European New Towns

Image, Identities, Future Perspectives

P.I.E. Peter Lang

Bruxelles · Bern · Berlin · Frankfurt am Main · New York · Oxford · Wien

Pascaline GABORIT

European New Towns

Image, Identities, Future Perspectives

This publication has been co-financed by the European Commission, Programme CIUDAD, Neighbourhood and Instrument partnership

The ENTP would like to thank for this publication: Andrew Johnson, President of the ENTP, leader of Harlow council; and the municipality of Zoetermeer for the financing and feedback on the first "New Towns study". We would also like to thank for the interviews: Jean-Pierre Marchetti, General Director of services, SAN Marne-la-Vallée-Val Maubuée, and Philippe Aumard, vice President of SAN; Jukka Kullberg, head of planning services, Vantaa; Gunilla Edwards, economic department, Basildon district Council.

Éditions scientifiques internationales
Brussels, 2010
1 avenue Maurice, B-1050 Brussels, Belgium
www.peterlang.com; info@peterlang.com

ISBN 978-90-5201-671-9
D/2010/5678/55

Library of Congress Cataloging-in-Publication Data
Gaborit, Pascaline. European new towns : image, identities, future perspectives / Pascaline Gaborit. p. cm. Includes bibliographical references. ISBN 978-90-5201-671-9 1. New towns—Europe. 2. City planning—Europe. I. Title. HT169.57.E84G33 2010 307.76'8094—dc22 2010032613

CIP also available from the British Library, GB

Bibliographic information published by "Die Deutsche Nationalbibliothek"

"Die Deutsche Nationalbibliothek" lists this publication in the "Deutsche Nationalbibliografie"; detailed bibliographic data is available on the Internet at <http://dnb.ddb.de>.

Contents

FOREWORD

New Towns: How to Draw Lessons from the Past to Better Anticipate the Future

By VINCENT FOUCHIER

A revival of national governments' renewed interest in the New Towns model is materializing across the world. The massive urbanization experienced by the Western world has driven many European countries to construct New Towns. But in Europe the current trends in population and economic growth no longer seem to justify the continued establishment of New Towns; instead, we wish to reinvent existing New Towns and draw lessons from them. It is assuredly strong economic growth that allows and sets forth the realization of New Towns; in a context of massive and rapid economic growth, existing cities and towns can no longer adequately respond to the take up of new inhabitants, and it becomes more effective to obtain natural land that can be developed in a coordinated manner. It is also the speed of economic growth that creates space, both financially and operationally, for the establishment of New Towns. We can observe this phenomenon happening across Asia (China, Malaysia, Hong Kong, etc.) or in the South Mediterranean.

This present period is therefore conducive to an exchange of experiences. However, there is currently no transferrable model, and it would be a mistake to assume that a solution in one context could be duplicated in another, either for the better or for worse! Each New Town must pursue its own set of innovative responses.

There is a great deal to learn in terms of success, but also in terms of failure or the approaches to overcoming challenges. We can take five fundamental lessons inspired by our European New Towns. The first key lesson is that New Towns are conducive to innovation, future aspirations and experimentation. Since many have been constructed in a short timespan, New Towns allow for an intellectual flowering that serves as a reference and window of opportunity.

They allow for exploration of new technology, forms, methods with positive outcomes (progress, comfort, originality, modernity), and sometimes, with negative consequences of experimentation. Many responses, as well as urban planning concepts and theory, have been introduced in New Towns and then spread in other urban areas; for example, we think of the link between public transportation and population density, water management, the design of public spaces and technological networks, financing of urban development, cultivating socially active communities, the role of art, etc.

But too much innovation runs the risk of compromising the project as a whole; some neighborhoods suffer the consequences of experimentation, leading to a domino effect where one or several experiments start to pose problems rather than offer solutions. Conversely, too much conformity leads to New Towns without souls, ordinary and scarcely attractive. It is above all a lesson from the past: we have witnessed the consequences of multiplication without end of "grands ensembles", or dormitory towns, towers or strips of housing, especially in Eastern Europe, but also in France in the 1950s. The homogeneous suburban areas in certain New Towns pose or will pose identical challenges. Therefore, it is necessary to strike a subtle balance between innovation, originality and commonality … often regulated by market forces; New Towns must attract a new clientele.

There is no urban planning without clientele! Without a shadow of a doubt, it is for this reason that the new renders the old obsolete in the very hearts of New Towns; the newest neighborhoods are always "the most beautiful", while the original neighborhoods lose their attractive qualities. This image issue is one of the major challenges that New Towns are currently facing. One challenge is the ageing of the built environment in New Towns (after 15-20 years) that is emerging in a big way along with this period of redevelopment and renovation; we need to cultivate original responses to this problem which is very specific to New Towns (unfortunately, in most cases, demolition is the preferred option, in the absence of planned alternative solutions).

This brings us to the second key lesson, bearing a great importance, all the more so in the face of challenges with sustainable development and environmental crises: the adaptation of urban planning.

New Towns should not be considered finished projects. A New Town must always remain new! It must continuously adapt to changes with respect to society, tastes, and lifestyles. Far too many New Towns never knew to or never could cultivate an evolving, adaptable and resilient urban fabric … and, therefore, were unable to respond to the changing needs of populations and of the socio-economic dimension. Long-term urban projects, architectural decisions, land use management issues and urban planning standards should be approached with a view to long-term flexibility.

The third key lesson, linked to the second, is that forecasting the future is not an exact or reliable science. Town planning requires making decisions about what is yet to come: is growth a perpetual phenomenon to be taken for granted? Will lifestyles evolve (etc.)? New Towns must demonstrate their adaptability in the face of changing contexts. Some New Town projects have suffered from grave challenges owing to the derailing of trends which had been predicted. There are a number of examples:

- Shannon (Ireland) should have been profiting from Transatlantic air traffic in the 1950s … but advancements in technological progress meant that airplanes no longer needed to stop in Shannon to refuel before crossing the Atlantic. The New Town currently has a population of 10,000 inhabitants.
- Le Vaudreuil (90 km from Paris) should have experienced economic growth upon the industrial development that flourished along the Seine River to the sea… but the manufacturing sector did not retain its strong growth of the 1945-1975 period (or "Thirty Glorious Years" as it is called in France). The New Town now has a population of only 3,500 inhabitants.
- Cumbernauld (Scotland) was planned in the 1950s in response to the economic development of Edinburgh and to mining activity… but this sector quickly dissolved. The town has 50,000 inhabitants today.
- The five New Towns surrounding Paris were the quintessential response to the babyboom phenomenon and to the "Thirty Glorious Years" in the Ile-de-France region. They were designed with the goal of accommodating 14 million inhabitants in the region by the year 2000 … but this forecasted figure had fallen to 11.5 million by 2010 and the New Towns have to eventually rethink their population goals on a general level.

A healthy dose of skepticism in urban planning is necessary; the future is not set in stone and the town must adapt to potentially very different situations.

The fourth key lesson has to do with the success or failure of New Towns in terms of their location and their link to the closest metropolis. Should they be urban centres in their own right – stand alone towns – or urban entities well-integrated into their metropolises? The case of London, where New Towns are located beyond the green belt (approximately 90 to 100 km from London), cannot be appropriately compared with that of the Parisian New Towns, which are located at the edge of the metropolitan region and treated as part of the job pool and Parisian metropolis. In both cases the results are conclusive, but also demonstrate the respective limits. Milton Keynes (90 km from London) represents, just as Marne-la-Vallée (25 km from Paris) does, the symbolic New Town of global success.

We can also cite the example of the New Towns near Amsterdam, which allow us to reflect on the location and temporal planning of European New Towns. The New Town of Lelystad was built on a vast polder 50 km northwest of Amsterdam at the end of the 1960s; 15 years later, another New Town – Almere – was created and sandwiched between Lelystad and Amsterdam. Almere has become a "boom town" (185,000 inhabitants at present and 350,000 projected), while Lelystad suffers from competition and is home to only 70,000 inhabitants.

The case of Marne-la-Vallée, along 28 km of the metro rail network connecting it to the centre of Paris, merits mention; this location has allowed for the New Town's progressive growth and expansion from Paris to the East, demonstrating a certain flexibility with chaotic cycles in the real estate market. The experience of New Towns in Hong Kong (Shatin, Yuen Long, Tsuen Wan, etc.) is rather similar.

Whatever the situation, New Towns must be considered in tandem with metropolitan centres that necessitated their creation in the first place. Long-term metropolitan planning is demonstratively an imperative tool and complementary to New Town planning.

This brings us to the fifth and last fundamental lesson: there is not one prescribed method that guarantees successful social cohesion in New Towns. But it is the primary concern of New Towns to be able to offer liveability and social opportunities for their inhabitants, visitors and users. New Towns cannot be evaluated simply on the basis of their population or employment opportunities; it is the social world – rooted in the diversity of populations and functions – that really matters … that must first be created and then nurtured with time. Injecting a little soul – something that could make the people of New Towns happy in their identity: that is the challenge that must be embraced by New Town creators and planners, be they from the public or private sector.

Vincent FOUCHIER[1]

Deputy Director General

Institut d'Aménagement et d'Urbanisme d'Île de France

vincent.fouchier@iau-idf.fr

1 Former secretary general of the *Programme interministériel d'histoire et d'évaluation des villes nouvelles françaises* – History and assessment interministerial programme of the French new towns.

Executive Summary

New Towns and satellite cities are created by governments based on a master plan approach to solve problems linked to rapid urbanization, demographic growth (i.e., housing needs) and rapid development. In Europe, New Towns refer to towns created after World War II to meet housing needs and to alleviate traffic congestion around main capital cities. The aim of this book is not only to conduct comparative research about European New Towns – depicting their current situation in terms of the main historical, administrative, financial, economic, social or environmental features and challenges more than 30 years after their creation – but also to identify different paths for adapting to current challenges and to address the fundamental issues of image and identity. In other words, this book will examine the capacity of current New Towns, in a time of demographic stagnation and economic crisis, to manage their transition to maturity (i.e., becoming "Mature Towns") – to be viewed as pilot and liveable towns, rather than "dormitory towns".

Firstly, this book will emphasize the history of New Towns in Europe and several other countries. Established during the post-World War II period around major cities, their aim was to provide spatial solutions to population growth, alleviating the congestion of metropolitan areas. They were mostly born out of planned initiatives led by the state or in cooperation with the national administration. Specific policies and tools were put in place for their creation and to foster development. Yet, this pattern is undergoing radical changes across Europe; due to the drastic reduction of public spending, New Towns have to increasingly manage and adapt to new challenges on their own, especially in a time of financial crisis.

Secondly, the historical framework will conclude with an overview of the main challenges New Towns are facing in their age of "maturity". Thus, the book will focus on:

1. *Financial and administrative aspects*: Although the general overview seems quite negative in light of the State's gradual disengagement, it is important to examine the different national situations to better assess New Towns' needs and remaining financial resources.

2. *Current demographic trends*: New Towns had been experiencing strong population growth, which has significantly slowed in most cases.

The populations tend to be rather young, but New Towns are not spared from the ageing phenomenon and this challenge will require specific measures in the future. On the other hand, the situation across New Towns tends to vary considerably, as large towns continue to experience growth.

3. *Provision of public services and amenities*: As they emerge from the same historical pattern, all New Towns share the problem of providing public services and amenities. Basic facilities are in great need of attention, especially for rapidly growing towns, and "soft" investments are becoming increasingly important in fostering social cohesion and improving town image.

4. *Distance and transportation:* New Towns are typically well connected to their "mother cities" (e.g., London, Paris, Rotterdam, The Hague). However, local transportation, which involves connecting areas *within* New Towns, needs to be enhanced. Until recently, the reliance on cars featured prominently in urban planning. Improved public transportation systems could reinforce social cohesion and the use of local public services.

5. *The Rural – Urban Dimension, Management of Landscapes and Preservation*: City limits, borders, centres and peripheries, etc. Today, these concepts seem less and less relevant for urban planners. One of the major challenges is *polycentricism* and its territorial and social consequences, as well as the interactions between urban and rural areas. Most European New Towns are considered heirs of Ebenezer Howard's "Garden cities". Therefore, in a context of growing concerns about climate change adaptation and mitigation, these towns must develop sustainable landscape management methods to preserve their natural heritage.

6. *Housing*: Supported by national policies, large-scale housing developments are still undertaken in some places. The main problems are still present, including the lack of social diversity and the need for urban regeneration of housing stocks. These have a negative impact on the image of New Towns.

7. *Economic development*: Although it was not originally a priority, New Towns have had to fill the gap between them and other cities to become "complete" towns, not only dormitory towns. There are many obstacles to solving the issue, including difficulties with investment and business attraction, as well as the gap between jobs and skills.

8. *Social cohesion and the fight against exclusion*: Due to the lack of economic attractiveness, high unemployment rates, and deteriorating housing stocks, New Towns often face major problems with meeting social needs and addressing the resulting pattern of social exclusion. Solutions must be sought through several channels.

9. *Identity and sense of community*: The lack of shared identity and sense of community are fundamental problems for New Towns. This is especially so in a globalised world, where a town's identity can be an important tool for attracting investment and tourists and for promoting social cohesion. Therefore, special attention should be paid to this fundamental issue, for example, by fostering participation of the population and implementing branding strategies.

Introduction

Having observed for the last ten years European policies and actions promoting regional and urban development, it is my sense that the best way to start this book is by highlighting a very simple fact: Cities are now major stakeholders, not only at the local level, but also at the national, European and international levels. Moreover, cities are becoming partners of the European Union and its institutions. This is a relatively recent process, which is itself a corollary of the growing influence of regions and local authorities throughout Europe since the 1980s, emboldened by the creation of the EU Committee of the Regions and by the establishment of the first regional and local representation offices in Brussels.

On the one hand, as partners of the EU, cities participate in European projects within the framework of various European programmes, promote their interests through specific representative offices and networks located mainly in Brussels, and interface in thematic assemblies and groups. On the other hand, urban issues are increasingly taken into account at the European level[1], especially in the design of European tools, as in the framework of the EU Regional Policy instruments (since the EU does not have any competence concerning urban issues)[2]. This level of attention is in line with the increasing number of challenges cities face in terms of environmental issues, economic crises, pollution, social exclusion, or poverty.

Despite improvements in the recognition of cities' roles and policies, it does not always mean that sufficient attention is being paid to adapting to changes and new trends (in terms of lifestyles, better environmental policies, different commercial patterns, proximity and distance, towns' multifunctionality[3]).

---

1 Here are some examples of tools developed by the EU to foster sustainable urban development: Political declarations and charters (the Aalborg Charter, the ESDP – European Spatial Development Perspective, the Leipzig Charter); programmes and grants (INTERREG or URBACT); as well as technical documents (the Reference Framework on Urban issues for instance).

2 European Commission – Interservice Group on Urban Development, *Guide to the urban dimension in European Union policies*, Brussels, January 2010.

3 "La ville autrement", in *Alternatives Economiques*, Collection Poche, hors série No. 39, June 2009, chapters I and II.

It is also striking that towns are now perceived as "urban laboratories", permanently reinventing themselves in order to avoid the ageing of buildings and to adapt to the new trends and needs outlined earlier. But this dynamic also leads to competition between territories, whose survival depends on attracting more inhabitants, retail, and economic activities. It is about cultivating an image of the "attractive city" as opposed to the one of the "shrinking city". "Identity can mean any number of things for an urban area and the people that live and work there. It relates to tangible and intangible heritage: buildings, history, memories (...) Identity can also help to improve the image of an area, stopping a down cycle process, supporting social transformation by positively marketing a place"[4].

In this context of growing competition among European territories, there is no unique or homogeneous group of European cities. Urban areas differ in regard to characteristics and priorities, challenges and issues at stake, depending on size, locality and historical background. Dense town centres in major cities and other gentrified areas are perceived as "winners of the territorial competition" (characterized by high prices, different functionalities and proximities, etc.) whereas peripheries, suburbs, or peri-urban areas suffer from numerous problems.

New Towns find themselves somewhere between these two categories. They seem to be teetering at the edge of disadvantaged areas as well as at those areas where a more virtuous cycle of development are still possible. Therefore, this book does not aim to ponder the nature of European cities. Rather, the goal is to shed light on a specific category of medium-sized cities, often ignored, unrecognized, and cursed/prejudiced with a bad reputation. European New Towns are perceived as dormitory towns, places "without souls and without image". Yet, are towns intended for low urban density condemned to be "towns without substance" or, conversely, could these towns be considered "pilot towns" able to integrate principles of sustainability?

As underlined by the French researcher, Pierre Merlin, the concept of "New Towns" is ambiguous. There are neither time limits, nor geographic limits attached to this concept.

In an article published in the frame of the International Conference on New Towns in Tehran (2005)[5], Merlin identifies three categories of "New Towns": (1) New Towns built to host new capital cities; (2) New

---

4 Scheffler N., Kulikauskas P., Barreiro F., "Managing urban identities: aim or tool of urban regeneration? ", in URBACT Tribune, 2009, p. 9.

5 Merlin, P. "Les villes nouvelles à travers le monde – Essai de typologie", in International Conference on New Towns Report, Ministry of Housing and Urban Development of Iran, Tehran, 2005, p. 125-137.

Towns dedicated to the development of economic activities; and (3) New Towns built in reaction to the demographic changes and urban growth in major urban areas.

This book will focus on the third category of New Towns, which were established across Europe[6] in the post-World War II period. These new municipalities were primarily built around major capital cities, like London, Paris and Amsterdam, in order to prevent uncontrolled urban development, to answer the population growth in Europe, to alleviate the congestion of metropolitan areas and to achieve balanced territorial development at the national level[7]. According to Pierre Merlin, towns are considered New Towns if they meet the three following criterions:

(1) The building operation was thanks to a decision taken by national public authorities;
(2) The town was the result of a nationally planned initiative;
(3) Specific policies and tools were implemented for their creation and further development.

Cities destroyed in World War II needed to establish new housing areas to accommodate the population. As they attracted new inhabitants, many cities underwent a frenzy of anarchic sprawl in the form of satellite developments and polycentrism[8] – a phenomenon depicting New Towns. It is also interesting to note that this trend existed both in Western and Eastern Europe.

Urban design and conceptions about town centres for instances, were increasingly shaped by a global perspective, leading to a single urban style. What resulted was a levelling of urban design and an architecture that was similar in design all the world over[9]. Beginning with the Charter of Athens (1931), greatly influenced by the industrial mode of production (i.e., living, working, culture, shopping, traffic, etc.), town centres would be designated as shopping or administrative areas, rather than nodes of concentrated and diversified activities. In the period following the Second World – characterized by simultaneous population and construction booms – it became mainstream practice to divide towns into "mono-functional" areas[10]. This type of urban planning was a

---

6 In this book, the term "Europe" designates the current territory of the EU.

7 This definition of New Towns can be found on the website of the European New Towns Platform (ENTP): http://www.newtowns.net.

8 It was also during this period when New Towns were created in North West Europe, following master plans born out of this philosophy.

9 Krier R., "Composition of urban spaces", in *International Conference on New towns Report, Ministry of Housing and Urban Development of Iran*, Tehran, 2005, p. 477.

10 The role played by massive state investments (both in the Keynesian and Socialist systems and therefore in Western and Eastern Europe) encouraged bureaucratic or-

realization of "functional separation" and, therefore, promoted social disintegration.[11] Today, "whole generations are forced to live with the consequences of planning errors"[12] These errors have not only resulted from poor planning, but are then exacerbated in the economic downturn as the inner-city property boom eventually implodes and leaves stagnating, vacant and unsaleable properties in its path of destruction[13].

European New Towns are now experiencing radical changes and facing specific administrative and financial challenges. The national governments, which had been strongly involved in the creation as well as the administrative and financial management of New Towns, tend to disengage themselves. There are several reasons for this phenomenon, including drastic reductions in public expenditure and the present challenges faced by other European cities and towns (e.g., economic development in a globalised world, social cohesion and tensions, adaptation to climate change and environmental issues). European New Towns must carry on more independently, managing their transition from "New" towns to "Mature" and "Pilot" towns and adapting their responses to current and forthcoming global and local challenges.

In its contribution to the transition process, this book will draw a general picture of European New Towns. Situating them in a comparative perspective, it will identify their main features and challenges, highlight good practices, as well as identify challenges which need proper reactionary and adaptational measures for the long-term.

In this respect, the text will achieve its intended aims in two ways:

– On the one hand, it will use theoretical material to build an overview of the New Towns' original development policies and to identify common features; that is to say, an external vision offered by urban policy experts.

– On the other hand, practical elements and real-world examples will enrich the theoretical approach by materializing current and forthcoming challenges – an internal perspective emanating from the stakeholders themselves (e.g., urban planners, local leaders, etc.). These perspectives were collected during interviews.

In the first part, we will examine the historical processes and mechanisms which were used to create and manage New Towns, mainly in the United Kingdom, France and the Netherlands – countries which had

---

ganisation of the urban development in short periods led by multi-annual plans with an objective of rationality.

11 Or the absence of social relations.

12 Krier R., "Composition of urban spaces", in *International Conference on New towns Report*, Ministry of Housing and Urban Development of Iran, Tehran, 2005, p. 478.

13 See the development of Harlow's centre with empty retail offices.

developed similar policies in this field – but also in the Nordic and Central-Eastern European countries. To prevent any "Euro-centric" bias, we will also explore examples of New Towns in several so-called "emerging countries" in Asia (China, India, South Korea) and in North Africa (Egypt, Algeria, Morocco). These examples demonstrate the development of different urban policies in the creation of New or "Satellite" Towns throughout various world regions. Examining different approaches to New Town management presents a unique opportunity to identify good, innovative practices and to learn from other experiences. Attention will then turn to the following areas – repleat with challenges for these increasingly "Mature Towns":

- Finance and administration;
- Current demographic trends;
- (Under) provision of public services and amenities;
- Distance and transport;
- The urban-rural dimension, landscape management and preservation;
- Housing (ageing infrastructure and housing stock);
- Economic development;
- Social cohesion and the fight against social exclusion;
- Identity (city branding) and fostering a sense of community.

The methodology builds on empirical documents, municipal reports, theories from the fields of urban studies, environment and sociology, as well as the results and outcomes of European projects. In addition, interviews have been conducted with municipal technicians and politicians. Our case study will focus on town centres to illustrate the previous findings and to highlight good practices and practical elements. It should also be noted that "image and identity" are cross-cutting themes featured throughout t. his book.

1. The Creation of New Towns: a Brief History

In painting a picture of New Towns' developments and their different challenges, it is necessary to highlight the different approaches taken in their creation throughout Europe. Even if the aims were the same, each country resorted to its own tools and policies to support the development of new urban areas. These policies still highly impact New Towns' image and sense of community. Individuals, communities, businesses and authorities attribute different factors to the development of identity: history is one key factor, playing an important role in cultivating the bad image from which New Towns suffer.

For example, the New Town of Dimitrovgrad, Bulgaria, established on the 2nd of September, 1947, became a symbol of communist ideology. The town's conceptualization is reflected in its old slogan, "We build the town and the town builds us"[1]. Already on the eve of communism's fall in 1989, the town was experiencing a decline. Its industry was devastated, most of the factories closed their doors, and massive unemployment followed. Attempts to change the town's name to "Trakia" all failed. Inhabitants are now trying to differentiate themselves from their town's communist legacy, but the image is difficult to shake (the statute of Dimitrovgrad in the town centre still refers back to the communist times) and invites a lot of prejudice. Initiatives to introduce art galleries, tourism and other modern activities still fail[2]. Thus, a quick peek into the history of New Towns tells us that several constitutive factors play on both the internal and external image and identity of these municipalities.

1.1. General Historical Outline

The notion, "New Towns" can be rather ambiguous and difficult to pin down as a concept. Indeed, towns have been created steadily by almost every civilization throughout history, and many examples are

1 Vlady N., "Dimitrovgrad: quand l'identité cherche à s'émanciper du symbole", in *Regards sur l'Est, dossier sur les villes nouvelles*, November 2007.

2 *Ibidem*.

traceable to ancient times (e.g., Naples was Nea Polis[3]). Nevertheless, the global diffusion of the concept dates back to the early 20th century, when Ebenezer Howard developed the idea of Garden Cities in a book entitled Tomorrow, a peaceful path to Real Reform (1898). The book inspired several initiatives, such as the creation of the Garden Cities Associations, and the concept spread widely across Europe. It is also important to note that a New Town, or a planned community, is often defined in dictionaries or official documents as "a city, town or community that was designed from scratch and grew up more or less following the plan[4]".

Five different periods mark the development of the New Towns phenomenon in the industrialized world[5]:

- *From 1880 to 1920*, urban development was closely linked with industrial development, and spaces were shaped by several functions.
- *From 1920 to 1965*, suburbanization exploded with the construction of large housing estates and involved the strict separation of spaces. During this period, Garden Cities and New Towns were being created in the United Kingdom.
- *From 1965 to 1985*, population growth slowed, the construction of individual houses was on the rise and business activity began to relocate from towns to outlaying suburbs, consequently increasing the spatial polarization. The ideal status of garden cities was tested with the creation of New Towns around metropolises in Finland, the United Kingdom, the Netherlands, France, Sweden and Spain.
- *From 1985 to 2000*, spatial relationships within regions began to change as entertainment and cultural activities flourished in old town centres, gentrification crept into historic centres, town marketing and branding became popular, and the competition to become metropolises played out between towns[6]. From this period, territories – including New Towns – were searching for their own identity in order to better position themselves in a competitive market.

---

3 As underlined in the baseline study of the EGENIUS URBACT project, 2008 http://urbact.eu/thematic-poles/sustainable-urban-development/thematic-networks/egenius/baseline.html.

4 According to the statutes of the ENTP. See http://www.newtowns.net.

5 Gaborit P., "Development of New Towns and peri-urbanized areas", in *International Conference on New Towns Report*, Ministry of Housing and Urban Development of Iran, Tehran, 2005, p. 276.

6 The term "Metropolis" will be defined here as a big urban area characterised by its size, its high number of inhabitants, its influence as the economic, political and cultural centre of a country or a region. It is often considered an important hub for regional and international connections and communication.

• *From 2000 onward*, while the idea of massive housing developments has been largely questioned, some countries have designated potential growth areas for continued concentration of new housing activities. With a growing emphasis on "sustainable development,"[7] transportation/access, economic development, social cohesion and environmental protection must be taken into account and will be key policy issues.

The creation of New Towns was also initially linked to the concept of "peri-urbanization". Peri-urbanization is defined as "the process by which cities grow by increasing their suburbs[8]" and concerns most urban areas. For J.P Antoni[9], "peri-urbanization refers to a highly dynamic process in which rural areas, both close but also increasingly distant from city centres become urbanized". In most countries, peri-urbanization results from urban sprawl.

The American Chamber of Commerce considers it "a process whereby the outskirts of a town undergo a kind of piecemeal citification[10]".

New Towns emerge from the same processes as that of peri-urbanization – built especially to counter the extension of anarchic suburbs. In most of the cases, however, this objective was not achieved and New Towns did not prevent suburban sprawl. Born out of the same process as urban sprawl, they tend to encounter the same problems (social, economic and, to a lesser extent, urban), not as anarchically perhaps, but share in the lack of a sense of community. They grew in parallel with suburbs, though typically more at the periphery. Nevertheless, there are two major differences. Firstly, contrary to the development of suburbs, New Towns are based on government will and, therefore, walls were put up before people could forge town identity. The second main difference is the link to garden cities and a precursory approach of sustainability. Whereas suburbs are usually associated with the idea of unsustainable development, the planning approach of New

---

7 Following the World Commission on Environment and development, also known as the Brundtland Commission, the concept of "sustainable development" is here defined as development that "meets the needs of the present without compromising the ability of future generations to meet their own needs" – United Nations, *Report of the World Commission on Environment and Development*, 1987, General Assembly Resolution 42/187.

8 Webster Douglas R. *et al.*, "Emerging third stage peri-urbanization: functional specialization in the Hangzhou Peri-urban region", October 2003.

9 Antoni J.P., "Urban Sprawl as modeling: A methodological approach", in *Cybergeo*, Paris, No. 207, 1 March 1992, p. 14.

10 According to the article published on the PU-GEC (Peri-urbanization and Global Environmental Change) entitled "Peri-urban area". See: http://web.ntpu.edu.tw/~shuli/PUGEC_02Peri-urban.htm.

Towns integrated the concept of sustainability even before the term "sustainable development" was mainstreamed.

1.2. Why Build New Towns? Common Goals, Different Approaches

After World War II the baby boom phenomenon and a rapid industrialization entailed an explosion in urban development in European countries. The fast increase in urban population growth created problems for cities, including overcrowding and congestion. Thus, many countries decided to build New Towns on the periphery of those cities to better control urban growth, while also responding to the housing shortage and providing citizens with a better quality of life. New Towns were considered one way to alleviate the congestion in metropolitan areas and to achieve balanced territorial development at a regional as well as national level through the introduction of new "growth centres".

This is a logical segue to the concept of "Polycentricism", which has been a constitutive idea of urban management since the post-World War II period and the launch of the first European New Towns. Polycentricism is usually defined as a principle of organizationing a territory (e.g., city, region, country or even macro-region, as in the case of the EU) around several political, social, economic or financial centres. In other words, a polycentric city is characterised by the plurality of its centres. In the case of New Towns built at the periphery of big metropolis or capital cities, this notion is crucial since they were initially intended as growth centres for attracting people. Therefore, New Towns were tools for building polycentric urban areas at a regional level, as in the case of French New Towns around Paris[11]. However, New Towns were often conceived by urban planners as polycentric cities – made from scratch or by uniting several villages and boroughs – without any clear city centre. This polycentric characteristic of New Towns has resulted in major consequences, in terms of housing, transportation or social segregation. Polycentriscism is sometimes blamed for feeding New Towns' bad image because it weakens the New Town's identity, as compared with "old" cities which have identifiable historical centres and traditional characteristics. But, as will be illustrated in the second section, New Towns are finding ways to deal with the challenges of polycentricism.

Furthermore, New Towns conjured the dream of the "ideal city", since planners and politicians created them ex nihilo or, at least, were

---

11 Berroir S. *et al.*, "La contribution des villes nouvelles au polycentrisme francilien: l'exemple de la polarisation liée à l'emploi", in *Espaces et sociétés* 1/2005 No. 119, p. 113-133.

free to rearrange and develop old hubs in a more suitable way. One of the attributes of New Towns is their special bond with their natural environment, which often occupies a significant part of their territory (i.e., urban parks, green connections, etc.). They were associated with the "green cities" concept (until recent debates on the densification of cities versus urban sprawl). Currently, most New Towns are about 40 year-old, and the context in which they were planned has changed radically overtime.

The construction, planning and development of those new areas were based on the highest standards and architectural practices of the time, but now they must deal with new challenges, such as deteriorating and insufficient housing stocks, the regeneration of city centres, adaptation to new sustainable requirements[12] and so on.

How can they respond to new challenges, especially now that the disengagement of national governments and dwindling public funds seem to have become the general pattern in Europe? Indeed, most of the New Town initiatives were based on an interventionist approach on the part of national governments; New Towns once benefitted from various financial subsidies and incentives, while their administration was facilitated by dedicated agencies to support their growth. It is against this backdrop that we now turn to the exploration of why and how the main New Town policies were conceived in different European countries.

1.2.1. United Kingdom

Ebenezer Howard[13] was the father of the British New Towns movement; his book and the subsequent creation of his garden cities, Letchworth (from 1903) and Welwyn (from 1920), inspired planners and politicians. He brought attention to the problems plaguing unhealthy settlements in Victorian industrial cities and towns and raised awareness of environmental standards.

Nevertheless, the creation of New Towns required greater national government involvement in planning and development control, but this was not recognized until the 1940s.

---

12 Gaborit P. & Lamas Sanchez. P., *Contextual analysis of six New Towns*, ENTP, Brussels, 2006, NEWTASC INTERREG III B project. See: http://www.newtowns.net.

13 Ebenezer Howard (1850-1928) was a prominent British urban planner. He thought deeply about social issues, and one of his achievements was a book called *A Peaceful path to Real Reform* (1878) reprinted as *Garden cities of tomorrow* (1902). This book called for the creation of new suburban towns of limited size, planned in advance and surrounded by a permanent belt of agricultural land. Howard believed that such garden cities were the perfect blend of city and nature.

In 1946, the United Kingdom had to deal with the devastation of its urban landscape imposed by World War II; the destruction of housing was widespread and infrastructure had been disrupted[14]. The restoration of cities provided an opportunity for large scale improvement of urban conditions, and New Towns were an important part of this strategy. They were aimed at improving living and working conditions in both devastated areas and new settlements.

In 1944, Sir Patrick Abercrombie published a plan for Greater London, advocating the creation of 10 new "satellite towns", close enough to London to serve as "overspill" areas but also sufficiently remote to avoid becoming dormitory towns[15]. In the meantime, Lord Reith was appointed chairman of a New Towns Committee, which concluded that New Towns should be created by government-sponsored corporations financed by the exchequer. The Committee's conclusions resulted in the New Towns Act of 1946, providing the basis for New Town development in England. Stevenage was the first to be designed in October 1946 and it was intended to be a model of good practice.

British New Towns
First generation: Basildon, Bracknell, Crawley, Harlow, Hatfield, Hemel Hempstead, Stevenage, Welwyn Garden City, Corby, Cwmbran, Newton Aycliffe, Peterlee
Second generation: Redditch, Runcorn, Skelmersdale, Washington
Third generation: Milton Keynes, Northampton, Peterborough, Telford, Warrington

On the whole, 21 New Towns were created under the New Tows programme, mostly to relieve congestion in London and to reduce pressure for development in the green belt by moving people and jobs to new centres. In London, this would entail the planned relocation of some two million people; re-housing people on this scale meant finding alternative settlements because the existing urban areas were not considered acceptable[16].

---

14 As underlined by English Partnerships. See: http://www.englishpartnerships.co.uk/newtowns.htm.

15 Watson A.G., "New Towns in perspective in England", in *New Towns in Perspective*, INTA Press, 1991, p. 11.

16 Gaborit P. & Lamas Sanchez. P., *Contextual analysis of six New Towns*, ENTP, Brussels, 2006.

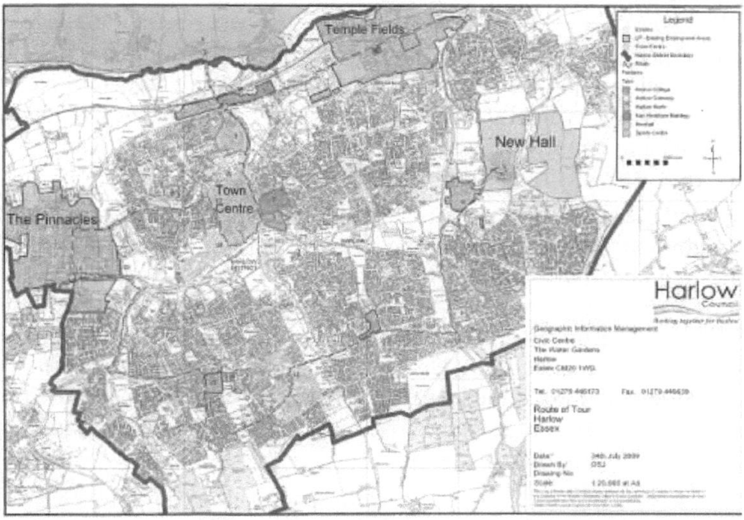

Current development of Harlow New Town

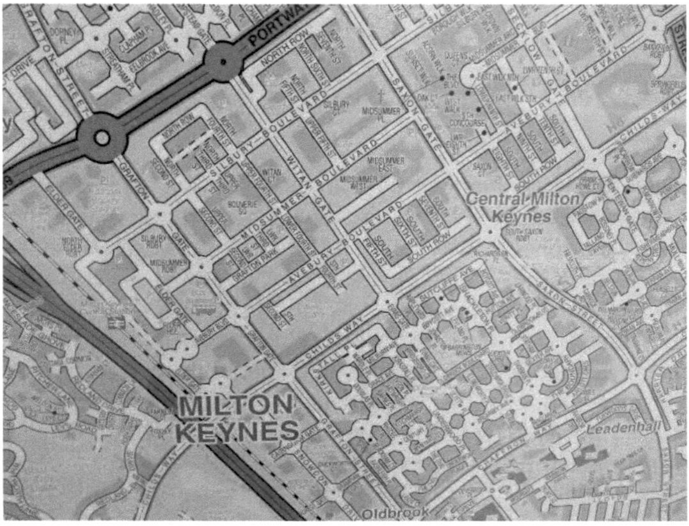

Map of Milton-Keynes and its grid-square design

1.2.2. France

In France, the creation of New Towns dates back to the 1970s, about one generation after the beginning of the British New Town movement. The latter were typically depicted as models to follow, and although there was no consensus among urban planners or politicians, Villes Nouvelles were clearly conceived in reference to New Towns[17].

Nonetheless, the concept of Garden Cities was adapted to the French urban context. Decolonization and massive rural migration had resulted in the uncontrollable growth of metropolitan cities, particularly in the case of Paris. New problems appeared, such as the lack of housing for newcomers, the under-equipment of Parisian outskirts and the transportation system's inadequacy. Unlike British New Towns, they were not designed as "self-contained entities". French New Towns had to integrate into the region to remain close to the mother city and connected to the regional transportation system.

French New Towns
Around Paris: Cergy-Pontoise, Evry, Marne-la-Vallée, Sénart, Saint-Quentin-en-Yvelines
Around other cities: Lille-Est, Le Vaudreuil (Rouen), L'Isle d'Abeau (Lyon), Rives de l'Etang de Berre (Marseille)

In 1965, a Master Plan for the Paris Region was adopted. It foresaw the creation of five New Towns in solving the region's urban planning problems. The concept was not fundamentally new, as Garden Cities had already been developed in the interwar period. The British New Town of Stevenage, for instance, was built in 1921 following the garden cities movement. However, the Master Plan went further as it took into account the economic, demographic and spatial growth of the region in its previsions. To end the dependence of Parisian suburbs on their mother city, it was decided to create both restructuring centres in the suburbs and new urban areas.

---

[17] Merlin P., "New Towns britanniques et Villes Nouvelles françaises: de la conception à l'achèvement", in *International Conference on New Towns Report*, Ministry of Housing and Urban Development of Iran, Tehran, 2005, p. 141.

The objectives of those new urban poles were to:

- Control urban growth;
- Concentrate urban development around well-equipped urban centres, which had to be supported by an efficient transportation network;
- Provide affordable housing;
- Balance jobs with housing (1 job for every active resident, the ideal situation being 1 local job per active resident).

Following the Paris Region's Master Plan, regional master plans were elaborated at the end of the 1960s, and four of them lead to the creation of New Towns. In total, nine New Towns were designed in the period between 1966 and 1973 mainly in the region of Paris.

View of Cergy Pontoise

Comparison of the New Towns' policies in France and the United Kingdom[18]
Similarities: • Method: deep involvement of the State and well-thought planning strategy • Guidelines: affordable housing to respond to growth, control of urban development, importance of quality of life and green spaces • Strategy: balance between jobs and housing to avoid creating "dormitory cities"
Differences: • Size: French New Towns relied on demographic perspectives that foresaw extreme growth, hence the need to create more important urban entities • Localization: British New Towns were conceived as "self-contained communities" and thus supposedly independent from big cities; on the contrary, French New Towns were located at the periphery of mother cities in order to attract more people and business, with a view to integrating into the existing regional urban structure • Housing: French New Towns chose shared housing as the main housing option • Management: British New Towns benefited from simpler and more efficient mechanisms

1.2.3. The Netherlands

In the Netherlands, the central government implemented the growth centres policy in the 1970s. Growth centres were municipalities designated by the central government to undergo considerable growth in the 1970s-1980s. Within approximately one decade (1975-1985), additional financial resources for housing and public services were allocated to these municipalities in order to compensate the cost of high growth[19].

Growth centres were intended to absorb the population growth in concentrated urban areas, following two core principles[20]:

- Efficient planning tool: Large cities were not able to cope with their growth, and thus, needed an efficient planning tool to help them;

---

18 Merlin P., "New Towns britanniques et Villes Nouvelles françaises: de la conception à l'achèvement", in *International Conference on New Towns Report*, Ministry of Housing and Urban Development of Iran, Tehran, 2005, p. 142.

19 Social and Cultural Planning Office of the Netherlands, *New Town Developments*, The Hague, March 2001, p. 3.

20 *Ibidem.*

- Well-structured urban poles: Some urban concentration would be necessary to prevent uncontrolled sprawl across the surrounding countryside; hence, the need for new, well-structured urban poles.

In the Netherlands, three different types of development patterns should be distinguished[21]:

1. New neighbourhoods, which were built in the 1950s-1960s in the immediate continuity of existing metropolitan regions. Such neighbourhoods were located within the limits of major cities and were designed according to local Master Plans. Their purpose was to provide mixed individual and shared housing along with a wide range of public services and public space. Buitenveldert, located south of Amsterdam, is a good example of this trend.

2. Accelerated growth of existing small or medium towns (groeikernen) to accommodate steady urban growth in the 1960s. The goals behind these operations were much more ambitious; the populations of these towns were expected to double within 20 years as they catered to the populations and activities of big cities in their vicinity. Purmerend, Hoorn and Alkmaar, which are located north of Amsterdam, are a few examples of this pattern.

3. New Towns created ex nihilo, benefitting from the status of groeikernen but constructed on a larger scale. This New Town trend included cities in the vicinity of Amsterdam, such as Almere and Lelystad (constructed after the IJsselmeerpolders' land reclamation), The Hague (Zoetermeer) and Utrecht (Nieuwegein). New Towns were not only intended to absorb the growth of big cities, but also to become growth poles on their own account; for instance, Almere has become the 7th largest city in the Netherlands with approximately 180,000 inhabitants.

---

21 Merlin P., "Les Villes Nouvelles à travers le monde: essai de typologie", in *International Conference on New Towns*, Tehran, Ministry of Housing and Urban Development of Iran, 2005, p. 134-135.

Leisure centre in Zoetermeer

City hall of Capelle aan den IJssel

1.2.4. Sweden

In many other European countries, operations were less ambitious (compared with the French, British or Dutch New Towns), as suggested by the Swedish case, particularly with regard to the development of Stockholm[22]. After World War II, Sweden's housing situation was in critical need of improvement, particularly in its major cities[23]. The

[22] *Ibidem*, p. 134.

[23] The abolishment of the *statare* system in 1944 was a major reason for urbanization. S*tatares* were agricultural laborers hired on the basis of one-year contracts. They received payment primarily in kind, a situation making it nearly impossible to leave. Upon the system's abolition, many workers who were no longer bound by yearly contracts went to the city to try their luck.

widespread housing shortage and low standard of housing called for a new policy.

A political consensus on how to deal with the problem emerged in the 1950s. Public construction companies would be in charge of building housing under the ownership of different municipalities. This idea was based on the goal of improving the standard of housing and, at the same time, clearing the slum settlements that existed in major cities. In line with this new policy, "city-like housing" communities were built on the outskirts of cities like Stockholm (e.g., Hässelby-Vällingby).

Meanwhile, plans were to avoid "dormitory cities" and, therefore, were focused on cultivating town centres, locating them close to the underground (metro) stations. Sense of community was the guiding principle behind the creation of such places. Communities were to foster well being and a sense of home, which the major cities could not provide. Swedish developments came to be called ABC-towns (arbete, bostad, centrum, i.e., work, housing, centre).

But these new suburbian developments were not enough to cope with the living situation in the 1950s. The Lex Bollmora law consequently allowed the cities' construction companies to build in other municipalities, paving the way for cheap housing construction in primarily new suburbs throughout the country.

The next step in the development of these New Towns and suburban settlements was the so called "Million Programme 1965-1974". The programme's aim was to eradicate the shortage of housing by building 100,000 apartments per year during the next ten years. This second phase focused on building housing, schools, hospitals, elderly retirement homes, swimming pools, kindergartens and communal houses linked to the settlements. These projects strengthened town centres, not only allowing people to work and live in the suburbs, but also to enjoy leisure activities and other necessary social services.

1.2.5. Denmark

Denmark's government, together with the Copenhagen municipality, decided to set up the Ørestad Development Corporation, based on the British New Towns concept. The corporation was charged with creating a New Town close to the Øresund bridge, which links Copenhagen with Malmö, Sweden.

The Øresund region – including the Copenhagen Metropolitan Area and Scania, the southernmost region of Sweden – is the new regional centre of Northern Europe[24]. The region benefits from its attractive

24 http://www.orestad.dk/index/uk_frontpage/uk_oresund_region.htm.

assets, including a highly developed infrastructure, which has enabled it to become the largest metropolitan area in Scandinavia boasting 3.6 millions inhabitants and a high concentration of knowledge and business activity.

Constructing a bridge between Copenhagen and Malmö presented a major action. On the one hand, Malmö is Sweden's third biggest city and is close to the prestigious Lund University. On the other hand, Copenhagen experiences steady growth and its central position in the region requires more space to cater to new jobs and residential areas. The bridge‘s addition enhances Ørestad's visibility, virtually turning it into the Øresund region's geographic centre; its proximity to Copenhagen and the Copenhagen Airport makes it a gateway to the city and offers excellent international traffic connections. Furthermore, it provides ample space for new construction projects that could not have otherwise been accommodated in Copenhagen's existing neighborhoods.

The reason for choosing the New Town approach to planning was that Ørestad would be an integrated part of central Copenhagen, rather than just another suburb. After Ørestad Development Corporation was founded, the ownership of the 310-hectare Ørestad area was transferred to the Development Corporation by the city of Copenhagen and the Danish government. The corporation's task was to develop and sell sites in the new quarter, to build the metro and invite tenders for operating it. Thus, the development of Ørestad and the sales of its sites have partly financed investment in the metro, which is inextricably linked to the project.

Ørestad, which will be developed over the next 20 years, is expected to function as a testing ground for new ideas[25]. The project's aim is to create a planned town respecting high architectural and environmental standards. For instance, one third of Ørestad consists of parks and green areas. The areas will not only provide a wide range of dwellings, but also many new businesses, so that Ørestad can be an active and lively town, instead of a typical dormitory town. A good example of this is the opening of Field's in 2004. Now the largest shopping and entertainment centre in Scandinavia, Field's has created many job opportunities, especially for those Swedes for whom it offers a closer place to work than Copenhagen.

1.2.6. Finland

Finnish New Towns are not part of a comprehensive national policy. However, they have played an important role in the history of Finnish

[25] http://www.orestad.dk/index/uk_frontpage.htm.

city planning[26]. This is particularly so for the Helsinki metropolitan region, which experienced very fast growth after World War II and, consequently, an accompanying housing deficit and an acute need to reorganize regional settlement structures. The Finnish situation was also marked by population displacements resulting from the USSR's occupations; people were relocated either to the countryside and or to cities.

Criticism of poor urban living conditions was already vivid in 1946, when Heikki von Hertzen published a pamphlet called "Home or Barracks for our Children"[27]. In this pamphlet, he denounced dense urban settlements in Helsinki and proposed to implement garden cities hosting one-family dwellings instead. According to von Hertzen's thinking, homes were to provide people with a greater opportunity to devote their time to their families, avoiding the seduction of urban culture (e.g., cafés, movies). He subsequently founded an organization called Housing Fund, which purchased land from the owners of Hagalund (one of the manors in the region), and organized the financing and construction of a New Town in the 1950s. In doing so, he combined his ideal of green spaces and his desire to enhance avant-garde design. The New Town would be called Tapiola[28], and it attracted the worldwide attention of architects and urban planners who saw it as the symbol of the rising industrial and welfare state.

Although Tapiola became a New Town, in so far as it included both housing and jobs, it never became an administrative entity in its own right[29].

Established just outside the municipal borders of Helsinki, it became a part of Espoo, the second largest city in Finland. Now, despite rapid urbanization, the Finnish mentality towards urban dwellings has not changed much; most Finns prefer one-family housing in the vicinity of nature. This desire explains the considerable growth of Espoo or Vantaa (4th largest city). Moreover, current projects such as the garden city of Suurpelto (Espoo) – the expected "never-seen-before entity"[30] – show that the interest in garden cities will not fade anytime soon.

---

26 Lapintie K., "New Towns, suburbs and urban sprawl – Struggling for integrity in the Helsinki metropolitan region", in *International Conference on New Towns Report*, Ministry of Housing and Urban Development of Iran, Tehran, 2005, p. 315.

27 *Ibidem*, p. 316.

28 The name *Tapiola* is a direct reference to *Tapio*, who is the Forest God in Finnish mythology. Right from the beginning, Tapiola's identity was based on features related to nature and the countryside.

29 *Ibidem*, p. 317.

30 http://www.suurpelto.fi/index.php.

School and Childcare in Vantaa New Town

Delegation of the European New Towns Platform in Vantaa

1.2.7. Central and Eastern Europe

It is not surprising that Central and Eastern European countries have experienced a planned approach to urbanization through communist policies[31]. This phenomenon can be seen in the creation of New Towns, which were mostly aimed at fostering a country's industrial development and simultaneously enabling urban planning. The number of industrial New Towns in the ex-USSR is estimated at 1100[32]. Their conception and creation reflected rigid planning principles, especially with regard to spatial repartition of housing, utilities and services. More modern housing provision was based almost exclusively on large housing estates inspired by developments in Western countries (e.g., France) during the 1950s-1960s.

After World War II, the Soviet industrial New Towns model was implemented in Central and Eastern European countries[33]. This was particularly the case around major heavy industry and mining areas, which were seen to counterbalance the pressure placed on capital regions. In Poland, for instance, Nowa Huta was founded in 1949 close to Krakow and to new steel plants; Nowe Tychy emerged at the beginning of the 1960s close to Katowice and was linked to a car factory[34]. In Hungary, the largest New Town, Dunaújváros, was built in the vicinity of steel plants. As on other continents, the Eastern European New Towns reflect the history and ideology of these countries in the 20th century. They can also be studied as "places for memory of urban contemporary history[35]". The creation of these New Towns' demanded enormous financial investments for infrastructure, facilities, energy provision, as well as the capital to build in newly designated areas.

Such territories are scarred with ruptures reflecting the different periods of pre-communism, communism and post-communist transition in only one century. As on other continents, these towns were meant to embody principles of pragmatism (e.g., economy, production) and utopia. The fact that so many New Towns were built can also be explained by several other factors:

- The tradition of state centralized planning;
- The necessity to rebuild after World War II;
- The idea of towns embodying communist ideology.

---

31 Gaborit P., "Development of New Towns and peri-urbanized areas", in *International Conference on New Towns Report*, Ministry of Housing and Urban Development of Iran, Tehran, 2005, p. 276.

32 Merlin P. "Les Villes Nouvelles à travers le monde: essai de typologie", in *International Conference on New Towns Report*, Ministry of Housing and Urban Development of Iran, Tehran, 2005, p. 128.

33 *Ibidem*, p. 129.

34 Ziobrowski Z., "Problems of new towns growth in Poland", in *Conference proceedings*, ENTP, Brussels, 2002, p. 2.

35 Bayou C and Le Bourhis C. "Les villes nouvelles de l'Est: quelles spécificités?", in *Regard sur l'Est, dossier sur les villes nouvelles*, November 2007, p. 1.

The Eastern New Towns were designed as urban laboratories[36]. Whereas Western European countries were influenced by the Garden Cities movement, endeavoring to build more "balanced cities" (with regard to jobs and housing, nature and buildings, self-sustainability and links to mother cities), communist countries established New Towns with specific economic and mono-functional features. However, some towns, like Eisehüttenstad, Eastern Germany (later called "Stalin Stadt") integrated architectural concepts glorifying communism. These were inspired by Soviet-style buildings, but also factored in gardens, green spaces and pedestrian areas. So, while motivations diverged along the iron curtain, some design aspects were commonly present in both Western Europe and the Eastern bloc.

Nowa Huta (Poland) is symptomatic of industrial New Towns created during the communist era (pictures by Eric Morency)

As far as the ageing of New Towns' housing and infrastructures is concerned, things are less comparable. The decline of industry in the Eastern Bloc strongly impacted the development of these New Towns. This phenomenon can be illustrated through the case of Poland's Stalowa Wola – "The town of Steel" close to the Ukrainian border. As the steel industry declined, the town became a "shrinking town", loosing inhabitants (especially young people) and forced to invest in new modern economic activities, like tourism. This trend is also apparent in Tatabanya, Hungary, where the traditional mining industry is overshadowed by a mix of new economic activities to guarantee the survival of the town.

The common feature shared by New Towns in Eastern Europe nowadays is their lack of identity. Indeed, the dominating concrete apartment blocks and the dispersed sub-standard services accompanying them fail to make for attractive urban environments. They were mainly

36 Bayou C and Le Bourhis C. "Les villes nouvelles de l'Est: quelles spécificités?", in *Regard sur l'Est, dossier sur les villes nouvelles*, November 2007, p. 6.

intended as monolithic places of work, less resistant to economic turbulence and offering only a weak capacity to self-regenerate. Following the collapse of communism, mining and heavy industries suffered from the economic downturn and serious economic problems prevailed in Central and Eastern European New Towns (e.g., unemployment shot up dramatically, forcing people to move to main cities and regional centres to find employment).

Consequently, there is an acute need for large-scale physical, economic and social regeneration of New Towns in these countries. However, these issues have not been yet properly or adequately addressed in Eastern European countries. Moreover, until recently, they lacked the necessary resources to carry out these tasks.

1.2.8. Peri-Urbanised Areas and New Towns in Emerging Countries

An increasing number of so-called "emerging countries" have recently chosen to build New Towns to deal with urban-related challenges, such as urban sprawl, pollution, provision of public services and infrastructure. In this section, we will also examine the issue of polycentricism, which does not merely concern the Western world, but also cities facing high population growth, like in Central and Eastern European, Southeast Asian and Arab countries.

Exported from British colonies with the proclamation "We want not only England but all parts of the Empire to be covered with Garden cities", the first modern New Towns arrived in developing countries in the early 20th century. Since 1945, New Towns have been used by post-colonial governments to achieve various aims. Perhaps, among the most impressive are new capital cities such as Brasilia or Astana in Central Asia. In response to problems in large cities, New Towns have been created around large cities such as Shanghai and Cairo. Many of these satelitte cities have been problematic despite the optimistic philosophy behind their establishment.

China

Unemployed rural workers in China are attracted to cities due to the mechanization of the agricultural sector and to growth of the industrial sector, but Chinese cities are not prepared to accommodate this population boom. Thus, since the 1980s, the central government has been launching several New Town projects (400 New Towns are expected to be built by 2020). China's urban population is expected to increase to 60% in 2020 (an increase of 15% compared to 2008).

However, these urban newcomers do not have enough resources to live in city centres, so they settle in outlaying areas. This results in a trend of segregation, which is directly responsible for the extension of cities, for the creation of new villages in industrialized zones, and for the emergence of "satellite towns". The pattern of polycentrism attracts two different social classes: the new migrants benefitting from basic services and the upper class who reside in luxurious communities. Yusuf and Nabeshima observe that: "The increase in China's urban population from 191 million in 1980 to 562 million in 2005 has called for massive investment in urban housing and infrastructure. Amazingly, China has been able to absorb more than 370 million people in its cities without the proliferation of urban slums, although sewage and waste disposal services have struggled to keep up with the demand. Between 1990 and 2000, 130 million new urban dwellers were provided access to improved sanitation facilities; nearly a third of urban population still lacks these services, however"[37].

The New Towns close to Shanghai are faced with such a situation. Shanghai has hosted ten satellite New Towns since 2001. These towns represent both recognition that an urban future means taking human needs, the environment and economic growth into account and the government's desire to control Shanghai's growth. The unique geographic position and proximity to Shanghai, as well as sophisticated port and railway transport, have facilitated the booming economy and population growth in these towns. However, their design quality still lags behind that of main cities. Some of those satellite New Towns reflect the side-effects of the Chinese government's centralised approach, which anticipates real needs rather than responding to them. The New Town of Lingang represents such a scenario.

While providing good public services, several universities, office and industrial spaces (especially maritime industries, thanks to its proximity to the new harbour of Shanghai), the people working there are still reluctant to live there due to the lack of shops, cultural infrastructure, industries and bad transportation system linking the city to the centre of Shanghai[38]. The New Town of Ordos in Northern China has a similar situation. Despite the government providing a two billion Euro construction budget, the New Town has difficulties attracting inhabi-

[37] Yusuf S & Nabeshima K., "Optimizing Urban Development", in China: Consequences, Strategies and Policies, World Bank Publications, January 2008, p. 8-9.

[38] Desné J., "Près de Shanghai, la ville nouvelle de Lingang attend ses résidents", Agence France Presse (AFP), 12th January 2009, published on the website *Aujourd'hui la Chine*: http://www.aujourdhuilachine.com/actualites-chine-pres-de-shanghai-la-ville-nouvelle-de-lingang-attend-ses-residents-10155.asp?1=1.

tants from the "old" town of Ordos, located 30 kilometers away from the New Town[39].

Low income groups prefer to live in the periphery with their families because rent prices are lower than in the centre. 120,000 inhabitants are now living in the Yangxing area of the Baoshan district in Shanghai, 80,000 of which are migrants. They rent provisional housing in informal settlements, whose existence is constantly threatened by the possibility that the municipality could decide to invest in the area and evict them. But these examples reflect only one "type" of New Town in the Shanghai region. Other New Towns are "privately" based and established to attract higher income groups. Some of these New Towns have incorporated a North European architectural style, like the town of Luodian which has a characteristically Scandinavian appearance.

Luodian, China

Egypt, Marroco, Algeria

In Egypt, a New Towns programme was initiated by Nasser in 1969. This has drawn from British and French experiences as well as Soviet planning models. With rapid urban population growth causing overcrowding in Egyptian cities, the government has been constructing around 19 New Towns away from the major centres of Cairo and Alexandria. Generous incentives have been available in attracting investment

39 Fried A. & Rambaud M., "Ordos, la ville fantôme", in *ARTE Info*, 5th January 2010.

to New Towns, including 10-year tax exemptions, cheap land and labor, and relatively little bureaucracy.

These New Towns are aimed at creating self-contained growth poles, adopting a voluntaristic planning approach, possibly as a way to control the dynamics between urban sprawl and natural preservation (e.g., using the desert to accommodate population growth).

The New Towns' ability to attract investors and residents has been determined largely by the success in reforming Egypt's economic policy framework and rehabilitating the human and physical infrastructures of these New Towns, in addition to improving the investment climate. However, the infrastructure has not kept up with the pace of development. So far, the authorities have largely failed to encourage workers to relocate; most workers continue to commute from their existing homes. But migration is expected to accelerate as pressures on existing centres increase, bringing with it new investment projects, particularly in housing and infrastructure[40].

Development plan for the New Town of Sidi Abdellah near Algiers, Algeria[41]

Moreover, the absorption capacity of New Towns (as estimated until the year 2000) poses a challenge to integrating all the new residents settling in outlaying desert areas. To satisfy the requirements of this

40 American Chamber of Commerce in Egypt, *Business Studies Series report on New Communities*, Cairo, 1995.

41 Ministry of urban planning and tourism in Algeria, *Study of the New Town of Sidi Abdellah*, 2009.

population, it was necessary to create a new urban framework based on standards and concepts that differ from those of New Towns. The concept of new settlements is based on self-help schemes, whereby approximately two million people will be provided assistance in creating desert towns, mostly on their own initiative. The primary goal is to encourage people to build for themselves, freely and legally. The basic planning concepts are to maximize advantages offered by New Town organizations and existing urbanization dynamics in the Greater Cairo region. In effect, this means that the proposed new urban entities (i.e., new settlements) will be self-sufficient communities, like New Towns, but located at the edge of the Greater Cairo region for the purpose of easily transplanting existing development dynamics in the region. The main difference between New Towns and new settlements in Egypt will be that new settlements will rely on private investments for their development and, therefore, will have to be located near existing residential and employment areas[42].

After its independence, Morroco experienced a rapid urban growth (due to migration from rural areas), making it necessary to expand cities and to provide more employment and housing solutions. In the next 20 years, the country needs to find ways to address high population growth, along with a 40% increase in the active population. This means responding to housing demand by creating 3,667,610 housing units and accompanying infrastructure (e.g., drinkable water, electricity, etc). Therefore, New Towns are one of the government's priorities for housing and urban planning. The Master plans call for the creation of several New Towns around Agadir, Marrakech and Tangier in the next 25 years[43].

For example, the construction of Ch'Rafate, a New Town close to Tangier, has been launched by King Mohammed VI. The New Town's ambitions are to host 150,000 inhabitants and offer 30,000 housing units in 2010.

The objectives are twofold: the first one is demographic, whereas the second one is about supporting the economic development of the TangerMed Harbour by housing the labour force working on this industrial site. The Master plan also includes development of public transportation, clean energy, green spaces, and a new economic business park.

---

42 Becard L., "New settlements: A new approach to solving existing housing and urban problems", in *The Expanding Metropolis: Coping with the Urban Growth of Cairo*, ed. Singapore: Concept Media/The Aga Khan Award for Architecture, 1985.

43 According to the project "New Medina – From Pilot Towns to Sustainable Towns: Reinventing New Towns" developed in the framework of the CIUDAD programme. See http://www.newtowns.net and http://www.ciudad-pilot.eu.

In Algeria, several towns are being built around the capital city of Tangier, like Sidi Abdelalh, built in 2003 and located 25 km west of the city. The same principles apply as in the creation of European New Towns after World War II (i.e., alleviating traffic congestion and absorbing housing demand in the capital city), but transportation and environmental concerns have been raised.

India

The experience of India has been different from other Asian countries because the creation of New Towns has played out in a context of democracy and less centralized urban development. Indeed, as underlined by Professor Bijlab Kanti Sengupta, "Since Independence in 1947, India has been engaged in promoting urban development"[44]. This is particularly the case in planned urban development, which has been inspired by the Garden Cities model imported by the UK. There were 118 New Towns in India in 1981, gathering 3% of the country's urban population. They were initially built to accommodate displaced persons after the 1947 partition, to host state capitals or to accompany the development of local industries (thus, "steel cities" like Bhilainagar, Bokaro or Durgapur). Moreover, like the European New Towns presented earlier, several Indian New Towns have mainly emerged near regional capital cities, like Naya Raipur (2001), New Chandigarh, Noci Mumbai, etc.

Despite these characteristics, "the role of New Towns in bringing a semblance of order to India's urban pattern has been minimal", as underlined by Professor Sengupta[45]. Indeed, urban development in India is often described as a polarized process between the development of major "megalopolises" and maintaining a high rural population. Despite the fact that several of the largest cities in the world are located in India, this country still has a relatively low urban development rate (29% in 2007), as well as a relatively low urban-rural migration rate (in comparison to other Asian countries)[46].

Nonetheless, urban development is still growing in India. Indeed, according to the national general planning commission, ten million new homes must be built each year until 2030 to meet the demands of new

---

44 Sengupta B. K., "Indian Experiences of New Towns Planning", in Merlin P. & Sudarkis M., *From Garden City to Urban Reconstruction: New Towns in Perspective*, INTA Press, 1991, p. 217.

45 *Ibidem*, p. 231.

46 Dupont V., "L'urbanisation en Inde: révision de quelques mythes et vrais défis", in *Rayonnement du CNRS* No. 47, Paris, mars 2008, p. 11.

urban populations and to improve living conditions in slums[47]. The recent construction of Naya Raipur, the capital of the state of Chhattisgarh (the 10th largest state), has involved new ideas with a special emphasis on medium density, pedestrian areas and sustainable development[48]. These publicly based New Towns should host five million inhabitants and contrast with other privately owned and established Indian New Towns, which function more like "gated communities"[49].

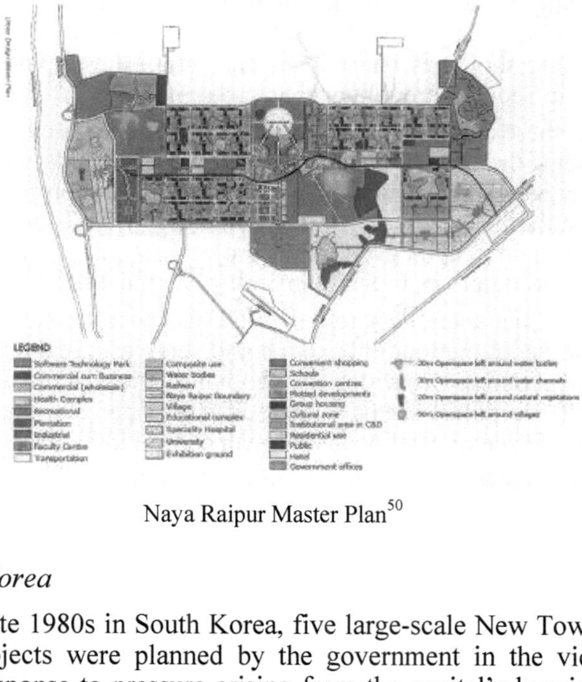

Naya Raipur Master Plan[50]

South Korea

In the late 1980s in South Korea, five large-scale New Town development projects were planned by the government in the vicinity of Seoul in response to pressure arising from the capital's housing shortage. They originated from the British New Towns concept. The total population of the five towns is estimated to increase to 1.2 million

47 Paquot T., "Éditorial – Villes indiennes", in *La Revue de l'Urbanisme* No. 355, July – August 2007.

48 Sudeshna C., "Naya Raipur, Chhattisgarh: A sustainable urban perspective", in *Sustainable solutions review*, p. 63-64.

49 Interview of Michel Sabatier, secretary general, Euro-India urban forum, Goa conference, 2007.

50 In Chatterjeed Sudeshna "Naya Raipur, Chhattisgarh: A sustainable urban perspective", in Sustainable solutions review, p. 63-64.

within a total area of 294 km^2. These New Towns will be within 20 km, a one-hour's commute, from Seoul's central business district.

These New Towns integrated fully equipped houses, commercial offices and shops, government offices, parks and recreational facilities, athletics facilities, green areas, and other conveniences. Accessibility to Seoul will be enhanced by the building of an intra-regional transportation network and mass transit system (based on electric rail) between Seoul and the satellite cities.

Following the success of the initial five New Towns, the South Korean government has decided in 2000 to construct another five to resolve housing shortages and unplanned, unruly development that continued in the Seoul metropolitan area.

Unlike the existing satellite cities, which are focused on industrial development, the New Towns will be targeted towards tertiary industries. Many have doubts about this ambitious housing project's potential to improve the lives of the poor and homeless, while some government economists claim that the New Town projects are mainly responsible for lowering housing prices in Seoul in both 1991 and 1992.

However, the actual debate lies in the fact that the new housing estates in the Seoul metropolitan area could serve as a real alternative for people and economic activities still willing to move into the area. Thus, this New Town policy seems to be a different approach than the previous policy of decentralisation and growth control in Seoul.

The New Town development seems to be a direct attack on the traditional beliefs of many South Korean planners. However, as mentioned before, under the market forces and economies of agglomeration, the Seoul metropolitan area attracts the majority of economic activities and migrants from the rural areas. Housing conditions in Seoul, especially for squatters in the over-idealistically named Moon Village, have become a social concern and a severe political problem. With planners still trying to decide on how to unite spatial balance and the effectiveness of "oversized" metropolitan regions, political needs seem to have triumphed in building New Towns in the Seoul metropolitan area. Thus, the New Town development debate continues in South Korea.

This overview helps us to understand that the issues dealt with and faced by New Towns are not only limited to Europe, but that the lessons taken from European New Towns can also help other countries.

2. New Towns Facing Challenges

The previous section, reviewing the historical context and the various approaches used to design New Towns in European and non-European countries, allowed us to identify the different policies and specific instruments used to develop New Towns throughout Europe. It seems relevant to turn the emphasis to the characteristic features of New Towns and to the current challenges they face. This analysis will be accompanied by a cross-cutting question: how do New Towns perceive themselves? Like towns demanding a specific approach, or as mainstream towns willing to forget their special history? Like independent cities or entities integrated into the development of a wider area?

2.1. Financing New Towns: an Overview

Financing New Towns has always been a highly debated and sensitive issue. It is a cross-cutting question echoed by each theme in the study. Should local resources come from state subsidies? Should finances come from the private sector or rather the local public sector?

In most countries, New Towns used to enjoy a special status and public funds (mainly from the national governments) to foster and support their development. They were mostly managed through specific bodies dedicated to their development, as could be seen in the United Kingdom with the so-called New Towns Development Corporations. But the situation has changed in recent years; New Towns are no longer considered "new", and the general pattern of disengagement on the part of national governments has led to major budget cuts and to the end of New Towns' special status. Increasingly, they have to manage by themselves, which raises important financial issues.

New Towns are severely in need of more funding, not only to continue their development, but also to renew and adapt themselves to current challenges (e.g., sustainability of buildings, economic competitiveness and so on).

Given the heavy investments they required in a relatively short period of time, it seems difficult for a New Town to be financially successful. However, in terms of management, New Towns often seem to act as laboratories (e.g. France and inter-municipal cooperation), enabling New Towns to host bolder projects better integrated into the surroundings.

2.1.1. Where Do Revenues Come From?

The financing of local authorities implies cooperation with central national governments. It involves funding with subsidies and other fiscal and financial devices that create interdependence. However, local fiscal autonomy is, to some extent, an illusion. Indeed, as underlined by a local civil servant working in a New Town: "There are no good local taxes for the simple reason that local resources never match local needs[1]".

France is divided into 36,000 units. The system uses a mix between local taxes and redistribution of taxes at the central level, with various adjustments and piecemeal arrangements. The main problem is the high number of small boroughs, which also explains why New Towns tried to merge several entities (i.e., through various administrative structures, such as the Syndicat d'Agglomération Nouvelle – SAN for instance, as will be shown later on). In the Netherlands, most revenues are nationalized; taxes are based on a national scale and the local expenditure results in a negotiation between the national government and each regional and local authority. The disadvantage of such a set up is its potential to feed a redistribution "free for all". In the United Kingdom, the centralized system showed political antagonism between centre and periphery. It has been increasingly centralized with the adoption of Acts which cover the whole territory, like the "Housing Act" in the 1980s, for example. The features included creation of regulatory committees and Commissions, as in the case of the Commission for New Towns.

2.1.2. New Town Administration

In France, the administration of New Towns was handled by the Ministry of Equipment. In each town, it was represented by a Committee of Public Planning called EPA (Établissement Public d'Aménagement in French). Through the EPA, the central state was responsible for:

- Determining the urban and economic orientation of New Towns;
- Investing in heavy infrastructure;
- Bearing the legal and financial liabilities of planning, which were defined in a national code of urban operation; and
- Grant subsidies to cover deficits incurred from the city's development and progressive tax raises induced by economic development (taxes paid by businesses).

1 Interview and training NEWTASC project 2006.

The local authority is represented by the Syndicate of Public Agglomeration (SAN), an inter-municipal system set up by representatives of municipalities to prevent the establishment of a unique commune.

At the time of their creation, it was difficult to include New Towns in the French centralized administrative system. All subsidies depended on different Départements (a bureaucratic level in the French system of local authorities, just below the regional level), making it difficult to operate a system where all subsidies would be received at the same time. French New Towns were encouraged to develop further following the successful example of British New Towns. In 1965, the central state planning document was approved by regional assemblies; and teams, involving national state representatives and people familiar with the local context, were created. The central state and the regions found funds to develop infrastructure, housing and services. An inter-ministerial group was created, and study and planning missions were sent to future New Towns. In order to keep land prices low, development plans were kept secret.

Between 1998 and 2004, most of the New Towns fell under regular statute. Since the EPAs have been suppressed, the central state has ceased to be in charge of planning and is now merely a partner of New Towns.

Marne-la-Vallée and Sénart, which still have great potential in terms of development, will change status in an upcoming, second phase. The concerned municipalities are free to keep their SAN status.

They are also entitled to take part in inter-municipal systems; for instance, Cergy-Pontoise has created a Communauté d'Agglomération which is an inter-municipal body.

In the Netherlands, New Towns are considered local towns and are managed by a local council like any other town. The financing of Almere has been an exception. New Towns were central government initiatives in the 1970s and 1980s. Over a period of approximately one decade (1975-1985), additional financial resources were allocated to these municipalities by way of compensation. Around 1985, the national growth centres policy was scaled down, and by the end of the 1990s, it had been replaced by the Vinex policy (as the population was still growing). A financial framework was established for these areas, and agreements with chosen municipalities were set regarding the use of this framework. Housing construction from the beginning of the century will remain considerable in New Towns, even though the growth centre policy has ended. A number of low status areas in New Towns fell under the Iadder status between 1991 and 2008 (Almere, Capelle and Spijkenisse). They will need additional funding to improve their social and physical situation.

In the United Kingdom, the Commission for New Towns, created in 1946 by the New Towns Act, was in charge of New Town administration. Land came under the ownership of New Towns Development Corporations (NTDC). In earlier New Towns, most of the land was developed for municipal housing estates. In later New Towns, more land was developed by private builders. As tenants gained rights to buy council houses and as the NTDC wound up, most of the land passed into private ownership. Land was bought at agricultural land prices and sold at residential land prices. Profits went to the public purse. Roads, services, parks and schools were also financed by the public, which created an equitable procedure.

In 1992, the NTDC was wound up and most assets and liabilities were transferred to the Commission for New Towns. This later merged with English partnerships which now own about 5700 hectares of land in the New Towns valued in excess of £1 billion. There are currently initiatives which would entail regional agencies of English Partnerships taking control of strategic sites and a New Towns Reinvestment Fund allocating profits from the sale of these sites.

2.1.3. Why Do New Towns Ask for More Financing?

New Towns were created to fulfil different functions, mainly to provide housing for the working class and to alleviate congestion in capital cities. After a period of robust public investments, the financing of New Towns has dwindled to a rapid withdrawal of public funding. This was especially evident in the case of the UK, where the New Towns Development Corporation was closed, but also in France, where the previous Public Development Institutions (EPA) were replaced by local councils (Communauté d'agglomération). Reactions varied across New Towns, but their development slowed mostly across the board. Although studies show that some New Towns are financially successful[2], the trends witnessed in the last couple of decades give the impression that the well being of current and future inhabitants has not been the guiding principle behind drastic decisions. As a consequence, the infrastructure in New Towns is still lagging behind (described in greater detail in succeeding chapters) and buildings are quickly deteriorating – all of which reinforce bad image.

2.1.3.1. Financial Justice (Equity)

From the beginning, there was a need to find land for housing and to create sustainable cities. Investment funds were scarce and homeowners

---

2 Peiser R.B. & Chang A.C., "Is it possible to build financially successful New Towns?", in *Urban Studies*, Vol. 36, No. 10, 1999, p. 1679-1703.

and developers found land designated for housing and development very expensive. Nowadays, needs have been transformed; some New Towns experience pockets of deprivation and housing estates are in need of major renovation. This is why a comprehensive investment programme is imperative if a spiral of decline is to be prevented.

2.1.3.2. Design Quality

Investing in quality, innovation (pilot actions) and quality of life (cycling paths, cultural events, etc.) should be a priority. Examples of public investments that make New Towns successful will be treated further in section 2.9 about town image and identity. But some actions merit mention here: the selection of Helsinki-Vantaa (Finland) as the next capital of urban design in 2012 will definitely give credit to the New Town of Vantaa and acknowledge the valuable outcomes of innovative design, in terms of architecture and planning[3]. The town of Cergy Pontoise (France) also scored points for Design Quality, winning a competition to host the new Archeological Department of the Louvre Museum. Vantaa and Cergy Pontoise serve as examples of how design quality will shape the future of New Town image and identity.

2.1.3.3. Sustainability

Considering current stakes, investing in sustainability – that is, environmental management, energy efficiency, green transportation, ecological corridors, water quality or green spaces – is now a major challenge for New Towns. But, as in the case of Design Quality, transforming the town from a peri-urbanized area to a liveable city requires public and private investments, as well as competition between territories. The New Towns better integrating environmental and sustainability concerns and better adapting to lifestyles may as well be winners of the transition from New Towns to mature cities.

Sustainability, however, should not be limited to construction, water and energy management, but should also pay mind to the evolution in lifestyle patterns. Projects to erect huge commercial centres with parking spaces in New Town centres may indeed be confronted with different lifestyle patterns and buying habits, underestimating the Internet's increasing role in satisfying consumer demand and the evolution towards softer modes of transportation (e.g., cycling, pedestrian). These trends have a strong potential to characterize future societies; New Towns and older cities may as well anticipate the need to incorporate such functionalities in their development projects.

---

3 See for instance the ENTP Newsletter March 2010 at http://www.newtowns.net/Newsletter%20March.pdf.

Finally, sustainability cannot be adequately addressed without allowing for inhabitant consultation. Serving as laboratories for urban development, New Towns had indeed been the dream of planners and architects. But this dream has had to face reality: the needs and desires of society were not sufficiently taken into account. It is a cause for consideration whether New Towns do not expose the limits of urban planning.

2.2. A Changing Population

Demographic changes have been recently recognized as one of the key issues for Europe's ageing population[4]. The main problems identified at global level are increasing life expectancy, growing health needs, gender issues, decreasing birth rates, and changing lifestyles[5]. However, the impact of changing demographic patterns on urban planning is a key issue for New Towns, a highly underestimated one, in fact. Local authorities not only need to plan for more amenities and services, like childcare, but also for changes in housing patterns, which themselves are subject to changes in the population. For example, an increasing number of multi-story dwellings need to integrate modern amenities, such as elevators, for the improved mobility of elderly people. But urban planning is also at the core of demographics because it needs to integrate all functional aspects of society, such as location of leisure activities, schools, and retail. Therefore, as lifestyle patterns mutate, urban planners need to transform the city once again in response to new needs, demands, and preferences[6].

2.2.1. Demographic Trends

As areas designed from scratch to accommodate the spill over from big cities, New Towns were conceived to absorb a rapid growth in line with the demographic developments of the post-World War II period. At the time, several phenomena combined to reinforce the predominance of urban development: the baby boom, the rural exodus and the arrival of many immigrants in Western European countries. In this context, New Towns have managed to attract a high number of inhabitants in relatively short periods of time. Population growth in New Towns is mainly due to inward migration, a trend that proves more prevalent in these municipalities than in "ordinary" medium-sized towns. What's more, most migration involves families[7].

---

4 Walker and Waltby, *Ageing Europe*, Open University Press, 1997.

5 Gaborit P. (ed). *Temps sociaux et parentés*, l'Harmattan 2008.

6 The EGENIUS URBACT project (2008) aimed at studying the changes of lifestyles on the planning of New Towns' centres: http://www.newtowns.net.

7 Social and Cultural Planning Office of the Netherlands, *op. cit.*, p. 5.

The capacity to attract inhabitants can be attributed to several factors. Firstly, New Towns provided a large and affordable housing stock compared to major cities that were no longer able to cater to sharp increases in housing demand. Furthermore, New Towns were planned around good transportation systems, linking them efficiently and rapidly to the neighboring major city. This in itself can be considered a driving force behind demographic growth. Finally, the enhancement of green spaces and of quality of life was an asset at a time when big cities were plagued with considerable urban problems (e.g., congestion and insalubrity).

The conception of New Towns was principally oriented towards a specific model of demographic development – that of the nuclear family (i.e., two parents and two children on average). In the meantime, population growth was expected to remain fairly steady[8]. However, urban growth slowed down drastically from the 1970s due to decreasing birth rates and a considerable slowdown in immigration resulting from the economic crisis or a near end to the rural exodus (which also meant the beginning of peri-urban diffusion). As traditional family dynamics changed (i.e., increases in divorce rates and single-parent families), the existing urban planning model became obsolete. For instance, in the Ile-de-France region, the government in 1965 had forecasted an overall population of 15 million by the end of the century; in fact, the administrative region of Paris had hardly reached 11 million in 2000. This major decrease in the forecasted figures had an impact on New Towns, in so far as the time required to pay off infrastructure was much longer than had been expected[9].

On the whole, the rate of inhabitants moving to New Towns has tended to decrease in recent years. This pattern is consistent with national trends, but is also due to other factors.

For one, some New Towns have reached their spatial limits and cannot expand any further; this is particularly so for towns highly dependent on space, such as Capelle aan den Ijssel in the Netherlands. Capelle experienced a rapid growth in the last 30 years, but its limits were reached in 2005. Consequently, there will no longer be anymore "great" development projects; rather, the emphasis will turn to management and

---

8 Hobson J., "New Towns, the modernist planning project and social justice: the case of Milton Keynes UK and 6th October Egypt", University Coll, London, Development Planning Unit, 1999 and Gaborit P., "Development of New Towns and peri-urbanized areas", in *International Conference on New Towns Report*, Ministry of Housing and Urban Development of Iran, Tehran, 2005, p. 276.

9 Burgel G., "La Ville Nouvelle comme politique urbaine: l'expérience des Villes Nouvelles de la région parisienne", in *International Conference on New Towns Report*, Ministry of Housing and Urban Development of Iran, Tehran, 2005, p. 257.

adaptation[10]. Yet, the main reason New Towns have experienced a slowdown in newcomers is the prevalence of severe problems hindering their development. As a consequence, there exists an acute need for modernization and/or renewal of derelict areas.

Nevertheless, European New Towns do not constitute a homogeneous group with regard to demographic development. In fact, trends can vary considerably across cases. While some New Towns have a declining population, especially if their whole economy was based on a single natural resource or industry affected by the 1970s crisis, others are still expanding and growing. The following chart summarizes and classifies the different situations.

- **Decline**	= **Stagnation**	+ **Slight increase**	++ **Strong increase**
New Town suffers from a shrinking phenomenon, i.e., its population is leaving (especially the young).	New Town no longer attracts many newcomers but manages to retain its population.	New Town keeps attracting inhabitants but at a slower rate than before.	New Town still experiences an important growth and has significant development projects.
Ex: Hoyerswerda (ex DDR)	Ex: Torfaen (United Kingdom)	Ex: Stevenage (United Kingdom),	Ex: Almere (Netherlands), Milton Keynes, Ashford, Basildon (United Kingdom), Val Maubuée, Sénart

Given current trends, the demographic justification of New Towns is much less applicable than it used to be. Obviously, objectives had been set according to the demographic perspectives at the time and they have been adjusted to account for major evolutions. The fact remains that the need for further urbanization has sharply decreased and that resorting to New Towns as growth centres is no longer the dominant approach.

Even though New Towns still fulfill their regional role in housing provision, priorities are now qualitative rather than quantitative[11].

---

10 Gaborit P. et Van Leuwen L., "Les Villes Nouvelles en Europe: de la ville pilote à la ville durable", in *Pouvoirs locaux* No. 60, 2004.

11 Fouchier V., *La politique des villes nouvelles (1965-2000)*, Programme interministériel d'histoire et d'évaluation des villes nouvelles françaises, Paris, 2000.

2.2.2. Population

New Towns have a specific demographic pattern compared to "ordinary" medium-sized towns, especially in France, the United Kingdom and the Netherlands. Their populations tend to be fairly young, and it is primarily young people who impact on the shifts in population. Inhabitants tend to be young couples with children, although this is less so the case now that children of the first generation of inhabitants have become adolescents or young adults. This trend was sometimes reinforced by national policies; for instance, the establishment of universities in French New Towns had shifted the population to a younger demographic.

A correlated feature is the relatively low proportion of senior citizens in New Towns. Yet, the ageing demographic phenomenon increasingly affects their lives. In the Dutch New Town of Spijkenisse, for example, only 10% of the population was above the age of 55 in 1983, as compared to 25% in 2009[12]. These demographic shifts require new policies and strategies, now and into the future, to provide the right type of housing and to deliver suitable transportation, other services and facilities[13].

One of the main challenges faced by New Towns is the need to attract higher income and highly skilled groups to guarantee social diversity and also to increase local resources. Indeed, research has shown that this socio-economic category is not drawn to New Town environments and prefers to live in neighbouring big cities or in small villages nearby.

Although many New Towns would like to have a more balanced demographic composition (in terms of age and socio-economic background) to match national or regional patterns, this goal has not been achieved in most cases. This situation partly explains why New Towns suffer from a negative image across Europe. However, studies in France and the Netherlands[14] have indicated that people who live in New Towns often wish to stay there and tend to move within the municipalities instead of leaving them, resulting in a sort of residential anchoring.

---

12 Statistisch jaaroverzicht 2004, Gemeente Spijkenisse, the Netherlands, 2004.

13 Gaborit P. et Van Leuwen L., *op. cit.*

14 Social and Cultural Planning Office of the Netherlands, *op. cit.*, p. 5 and Imbert C., "L'ancrage residentiel et familial en ville nouvelle", in *Note rapide Population et modes de vie* No. 376/D, IAURIF, Paris, 2005.

Case study: demographic evolutions in New Towns around Paris[15]

– In a context of continuing urban sprawl in the Ile-de-France region, New Towns still experience the strongest growth, especially Marne-la-Vallée. Furthermore, their weight in the regional population keeps increasing (6.8% in 1999).
– However, their growth rate has sharply decreased between 1990 and 1999. Demographic evolutions in the communes of the New Towns have been subdued and there has been no great increase or decrease in the populations. This constitutes a significant change compared to previous periods.
– It is also interesting to note that growth in New Towns are due less and less to net migrations; indeed, this rate was almost equal to 0 during the 1990-1999 period, whereas it amounted to +15,600 inhabitants during the 1982-1990 period. It is primarily their natural increase that enables them to keep growing (i.e. young households with children and few deaths, few elderly). But as this generation of young households becomes older, the natural increase will diminish, owing to a "mechanic" effect, along with the average size of households (since children will eventually leave home). Consequently, the migration rate will be less favorable unless the housing situation is quickly adjusted. It is not certain that in forthcoming years the natural increase will be able to compensate for a weak or negative migration rate.

2.3. Provision of Public Services and Amenities

Although growth and housing were declared priorities of New Town national policies, other concerns have rapidly emerged. Indeed, housing in itself was not sufficient to guarantee a good quality of life for the inhabitants. Inhabitants needed schools, hospitals and other public facilities. The question was: how to provide a well-equipped environment able to address various needs?

2.3.1. Fostering Image

Fostering a coherent, functional living environment is essential for a town or city. It is clearly a determining factor for businesses looking to set up shop and, thus, the competitiveness of a given territory[16]. Above all, it is a prerequisite for attracting and retaining inhabitants. This is particularly so for New Towns, which are often prone to a lack of social cohesion; owing to their fairly recent creation, people do not feel unified under a common identity. In this respect, public services (ranging

[15] Fouchier V., *La politique des villes nouvelles (1965-2000)*, programme interministériel d'histoire et d'évaluation des villes nouvelles françaises, Paris, 2000.

[16] Bouinot J., *La ville compétitive: les clés de la nouvelle gestion urbaine*, Paris, Economica, collection Gestion, 2002.

from education, health and social care to culture and leisure) are a crucial tool in preventing the "dormitory town" syndrome and in making places more liveable and attractive. In most cases, New Towns tend to function as residential communities, rather than communities with amenities. This is the reason why public services and cultural facilities represent an opportunity for New Towns to cultivate social cohesion and a sense of community belonging.

2.3.2. Basic Public Services and Amenities

It is generally agreed that New Towns have been striving to develop a wide range of basic public facilities in a relatively short period of time. However, the overwhelming conclusion is that New Towns are still lagging behind "ordinary" towns and that progress must be made in several fields.

To begin with, it makes sense to point out that the situation in Central and Eastern European New Towns stands in sharp contrast to that of Western New Towns[17]. Basic services (e.g., sewage, water supply and sanitation) were a top priority at an early stage in the British, French or Dutch New Towns' policies, while the list of necessary amenities grew shorter over time and taken for granted. In Stevenage (United Kingdom), for instance, the required facilities in 1951 were water supply and sewage; while in 1971 the requirement turned to hot water supply, and from 1991, central heating became the principal demand[18]. Despite their many years of existence, many Polish and Hungarian New Towns were still lacking basic services in 2009. For instance, 80% of households in Erd (Hungary) do not have access to drinking water, which makes it one of the major action items planned in the mayor's mandate[19]. It is therefore difficult to draw comparisons in regard to public services and amenities between Central and European New Towns and their Western counterparts.

Health care has often been a priority of New Towns. The table below shows the number of health care facilities identified in six New Towns with a fairly comparable number of inhabitants (between 150,000 and 228,000). Most of the New Towns seem to have sufficient health care facilities, and in the case of Cergy-Pontoise it is very noteworthy. A low level of health care facilities can be a problem if inhabitants are forced to receive health care and hospital treatment in surrounding big cities,

---

17 ENTP study visit to Erd June 2009: Explanations by the mayor about the new project of clean water and sewage facilities co-financed by Europe.

18 See in: http://www.visionofbritain.org.uk/data_theme_page.jsp?u_id=10088380&c_id=10001043&data_theme=T_HOUS/

19 ENTP, *Board Meeting report*, Erd, 2009.

rather than in their own town. Moreover, this situation can entail accessibility problems, particularly if the town covers a wide territory.

New Town	Number of health care facilities (hospitals, clinics, dispensaries)
Saint-Quentin (FR)	2
Cergy-Pontoise (FR)	5
Basildon (UK)	2
Milton Keynes (UK)	3
Almere (NL)	3
Haarlemmermeer (NL)	1

Health care has received special attention in countries like the Netherlands, where a majority of New Town inhabitants enjoy the possibility of receiving hospital treatment in their community[20]. Sweden is another advanced country in this regard; for instance, the New Town of Katrineholm has developed a "health square" that offers activities promoting healthier lifestyles, disease prevention and supporting rehabilitation. People are invited to participate in lifestyle groups or themed evenings, while parental support or help for quitting smoking are provided. On the whole, an increasing number of New Towns are investing in health, either to develop high added-value activities (e.g., Sénart and its cluster on cancer research) or to raise awareness among the inhabitants and to advocate lifestyle changes (e.g. Basildon, United Kingdom, where some projects of "Our Basildon" have been designed together with the NHS[21]).

Facilities for seniors tend to be less developed than in "ordinary" medium-sized towns[22], but this can be attributed to the relatively low percentage of people aged over 65 in New Towns (as discussed earlier). Nevertheless, if supplementary facilities are not built, the low average numbers of nursing homes and elderly housing communities could present problems against a backdrop of an ageing population. Indeed, the challenges posed by this demographic phenomenon will not escape New Towns. Villeneuve d'Ascq (France) has seized the opportunity to launch initiatives in favor of elderly and disabled people. The charter of "handi-welcoming" retailers is one of its most prominent actions in the

20 Social and Cultural Planning Office of the Netherlands, *op. cit.*, p. 8.

21 National Health System.

22 Social and Cultural Planning Office of the Netherlands, *op. cit.*, p. 7.

field; retailers who signed the charter pledged to improve, among other things, access to their shops and the readability of labels and prices.

Education is obviously a crucial aspect of public services and New Towns are often well supplied. However, problems have arisen due to fast growth, underscoring the need to allow for more efficient provision of education facilities. Thus, as explained by the Director of Services of Val Maubuée (France) in 2009, "making the public services cope with population growth was and, is still, a challenge". Val Maubeé has used prefabricated buildings for schools and many are still being used.

It should also be mentioned that situations tend to vary across countries and municipalities and depend on who is in charge of the education issue. In Sweden, for instance, municipalities are responsible for pre-schools, primary and secondary education. The New Town of Katrineholm invests almost 40% of its budget in education[23]. In any case, with the renewed importance of competitiveness and human capital, the emphasis is increasingly set on this question. Although education is an important factor in bolstering competitiveness, it should also be remembered that it plays an important role in creating social links and in integrating young people in their living environment.

As such, schools must interact positively with the "urban fabric" and be located in the heart of districts, rather than in isolated places; they must be a factor in enhancing social cohesion[24]. In addition, education can have a significant impact on the balance of age in New Towns. For instance, the presence of universities or post-secondary schools will have an impact on the overall age of the population and on public services.

Case study: New Towns and universities in France
– In France, New Town national policy was deeply linked with the establishment or relocation of universities. This decision was aimed at boosting the development of New Towns by giving them the opportunity to foster cooperation with businesses and to simultaneously create an integrated student area. One could refer to a "French approach", although Louvain-la-Neuve hosts the largest university in Belgium and some New Towns have highly competitive post-secondary schools.

23 Katrineholm funding resources, *Mobilization of local revenue sources for city development and service delivery*, Presentation in Cairo, 2009.

24 European New Towns Platform, *The top 8 specific challenges for social cohesion in New Towns*, 2005, http://www.newtowns.net/themes.

– **The Université 2000 plan** was implemented in 1990. Its objective was to adapt the French education system to the spike in student numbers and to operate qualitative and quantitative balancing (e.g., between Ile-de-France and other regions, or between big and smaller cities). The plan advocated the construction of 4 new universities in Marne-la-Vallée, Cergy-Pontoise, Saint-Quentin-en-Yvelines and Evry.

– **What was the impact of the Université 2000 plan on the development of New Towns?** Although it represents the opportunity to develop competitive clusters and attract businesses (e.g., Cité Descartes in Marne-la-Vallée), local authorities' expectations were unequally satisfied[25]. Usually, the logics behind academic and research developments diverge. For instance, the local level expects the results of applied research to be directly used by business, but this does not necessarily meet academic objectives of developing high level basic research. Furthermore, some New Towns complain that the large number of students living on their soil is an obstacle to attracting high-income groups (e.g., because students tend to prefer cheap housing and shopping facilities). Along with the current stabilization or (even) decrease of student numbers, the mounting financial constraints placed on local authorities explain why this subject is still a hot topic.

2.3.3. Increasing Focus on Entertainment

Currently, developing leisure activities is a fashionable trend in cities. Progress has to be made in this regard to improve New Towns' overall image in the eyes of companies and high-income populations. Nevertheless, many interesting projects have been implemented in recent years.

Cultural facilities are often insufficiently provided, which explains why New Town residents are less likely to attend cultural events than the average medium-sized town inhabitant[26]. Most of the time, New Towns do not provide comprehensive leisure activities, implying that their residents attend cultural events in the big city nearby[27]. This pattern contributes to a vicious circle, reinforcing the image of New Towns as dull places where nothing happens. Obviously, the point is not to compete with the big city, but rather, for New Towns to provide their own cultural activities and facilities, which could complement existing opportunities in the area. Culture can be a lever for social cohesion if local authorities approach it as a way to bring people together from different backgrounds and also if they strive to make it accessible.

---

[25] See: http://www.amue.fr/presentation/articles/article/une-evaluation-du-plan-universite-2000-1990-1995.

[26] Social and Cultural Planning Office of the Netherlands, *op. cit.*, p. 7.

[27] ENTP, *The top 8 specific challenges for social cohesion in New Towns*, 2005.

Culture is an efficient tool for building a town's distinct identity, especially at a time when towns and cities use it more and more for (re)branding purposes[28]. This is why, increasingly, New Towns want to play home to iconic cultural buildings.

Case study: Hergé Museum in Louvain-la-Neuve (Belgium)

– The president of Hergé Foundation, Fanny Rodwell decided to locate a museum dedicated to Hergé, the creator of Tintin comics, in Louvain-la-Neuve. The foundation paid for the entire building (15,000,000 €) provided that the municipality of Ottignies-Louvain-la-Neuve and its greater region, Walloon Brabant, pay for infrastructure (e.g., new parking spaces) and transport from Brussels. The museum is located on the edge of the town centre but is connected to this centre by a bridge. Cars can enter the museum area without passing through the car-free town.

– The museum opened in June 2009 with the **following aims and outcomes**:

- 200,000 visitors/year to the museum
- Increased interest in the town; more tourism
- Higher revenues from comic shops
- Higher revenues from the museum's restaurant/café
- Strengthened positive image and identity of the town and its citizens

– **What are the replication possibilities?** The creation of museums with a focus on internationally known artists/artistic styles could be replicated in other towns to enhance their image and identity. It is often a costly procedure and seldom pays off in economic terms. In the case of Louvain-la-Neuve, the museum was entirely paid for by a third party; this is a great advantage but can hardly be replicated elsewhere. Bidding for museums by offering to build the external facilities (i.e., infrastructure and parking) can be used in other cases, but town administrations should be aware that this is usually very costly. In any case, it should not be forgotten that the Hergé museum is closely linked to the Belgian identity, which has been partially shaped by its internationally renowned comics.

In and of itself, sport is often not considered a category in surveys. Yet, sport activity is as much a cultural tool in bridging the gap between people of different ages, cultures and classes. Indeed, New Towns have to put up with particular problems stemming from a strongly individualized social structure that often encourages households to keep to them-

[28] This is particularly the case for former industrial cities. See the case of Lille (France): since its designation as the European capital of culture in 2004, the city has managed to maintain an image of liveliness, thanks to major cultural events organized every 2 years.

selves. Consequently, traditional sports clubs tend to experience an ever weaker commitment by residents, even though they could present an opportunity to foster social cohesion. Sport is moreover a good way to address issues related to children, especially with regard to obesity, New Town populations sometimes experience severe overweight problems[29] because of financial difficulties experienced by households and the single-parent family pattern.

Therefore, building iconic sports facilities can be an asset for New Towns because it can enable a great number of residents to engage in sports activities while improving the quality of facilities. In any case, it is crucial to think first and foremost about the potential benefits to the community, e.g., by ensuring that facilities are physically and financially accessible to all. The Stedenwijk area in Almere (the Netherlands) is home to predominately low-income households, with 45% of the population of ethnic minority background, numerous single-parent families and with only 8% able to afford enrolment fees for sports clubs and activities. In light of this context, it is important to enable accessibility for the largest number of residents, i.e., facilities must be established not only in terms of the added value they bring for the town or the whole region, but also with a view to improving residents' quality of life. Zoetermeer (the Netherlands) has also developed facilities to promote sports among inhabitants, even if most of the initiatives are privately led (e.g., indoor skiing facilities called "Snowworld", a whitewater sports complex called "Dutch Water Dreams" and a mountain bike and Formula Two racing events). This promotion of sport offers alternative activities for inhabitants while improving their quality of life.

Case study: Harlow Gateway project (United Kingdom)
– Harlow Gateway is part of Harlow's regeneration strategy. It is a £50 million project backed by the Department of Communities and Local Government, Harlow Council, Harlow Sports Trust and English Partnerships. It aims to deliver new sporting facilities alongside much-needed housing for the New Town. – A £1million **athletics track**, a £3 million **football stadium** and a £25m state-of-the-art **community leisure centre** (with a swimming pool, a sports hall that can double as a conference/concert/exhibition venue) have been built with of the goals of: • Generating a major boost for sports, leisure and lifestyle activities • Responding to the housing shortage

29 In the case of the city of Almere (the Netherlands) for instance, 6,000 children are overweight.

• Improving the New Town's overall image
– The project combines **2 crucial dimensions**: (1) its use by competing athletes (i.e., the leisure centre has been earmarked as a training facility for Paralympians and Olympians in the run up to the 2012 London Olympics) and (2) its use by schools and the general community (i.e., the athletics track is the focus of a major training programme for schoolchildren in the town). It is interesting to note that Harlow intends to use these new facilities in the framework of the Olympic Games. Indeed, the town is one of just 65 major venues in the East of England and one of 600 across the UK to feature in the Olympic guide and one of just 10 Eastern area venues included in the Paralympic guide.
– **What are the replication possibilities?** Although a costly project, Harlow Gateway is based on an efficient partnership mixing the public and private sectors and has the potential to conciliate growth and urban renewal. It can be recommended provided that the project:
• is integrated into and coherent with a wider regeneration strategy (e.g., Harlow 2020);
• responds to an acute need/demand (e.g., outdated infrastructure, lack of facilities, etc.);
• targets the community as a whole via a long term vision;
• A profitable partnership can be formed to guarantee the best development possible for the project.

Shopping facilities have been insufficient in New Towns. They typically do not offer a central location and lack in leisure activities, which explains why attractive shopping areas were mostly found in neighboring major cities, thus reinforcing the "dormitory" image of New Towns. A great deal of progress was made in the 1990s with the construction of new shopping centres, particularly in Dutch New Towns[30], but it is still not enough. The relative number of people who shop in their place of residence remains less compared with other medium-sized towns. This also depends on the distance between New Towns and their mother cities. In the United Kingdom, for instance, New Towns were conceived as supposedly independent, "self-contained communities", which called for shopping facilities being taken into account in town planning.

In any case, shopping centres seem to lack in their attractiveness, as they have been criticized for not providing a pleasant atmosphere and for being poorly accessible by car[31]. Consequently, many New Towns are now investing to renovate those facilities; for example, the V2 shopping mall in Villeneuve d'Ascq (France) was completely renovated in 2004 and subsequently labeled "fluid, bright and warm". For highly competitive areas, the stakes are greater but essentially the same; shop-

30 Social and Cultural Planning Office of the Netherlands, *op. cit.*, p. 8.

31 *Ibidem.*

ping malls are designed to be national top-ranking facilities able to attract people and businesses. However, this "shopping mall" approach inspired by the United-States is criticized in towns such as Milton-Keynes or Sénart, where the shopping mall squeezes out other types of retail. To remedy this phenomenon, New Towns such as Louvain-la-Neuve are proposing to stimulate local retail to counter the attraction of shopping malls[32].

Vantaa's shopping centre: the second largest centre in Finland

Case study: Renovation of Vällingby Centrum shopping area in Hässelby-Vällingby (Sweden)

As the New Town of Hässelby started to take shape, a shopping centre linking to it was built. During its initial years in the 1950s it was extremely popular and hip. Eventually after several years, the area, without undergoing any renovation, turned into an obsolete shopping zone. **The renovation of the shopping area finally took place from 2003 to 2008**. A solution combining the preservation of the old with the addition of something new was found; it involved constructing a floating glass roof over the shopping street and building an additional mall.

– **The goals of the renovation and extension were to:**
- Promote the New Town's attractiveness as a place to live
- Improve the overall image
- Create jobs and enhance the economy

– **What are the replication possibilities?** The shopping area's revitalization was inspired by the fact the Kista Gallery, a nearby shopping mall which had been previously renovated, was doing very well. Nevertheless, the regeneration project always kept in mind the preservation of the town. The replication might not work in economic terms since it has been extremely costly. Nevertheless, it is a great example of how towns can preserve old areas and renew their image at the same time.

32 Interview elected representative from Louvain-la-Neuve 5th July 2009.

2.3.4. Main Problems Encountered

Money is Scarce

One of the main reasons why Dutch New Towns are lagging behind in public facilities is the central government's implementation of financial cutbacks[33]. Indeed, the growth centres policy offered financial compensation to New Towns in exchange for the construction of new homes and public facilities, which enabled the establishment of facilities, like sports centres. However, once compensation allocations were no longer granted, it was difficult for New Towns to continue such operations and to adapt to rapid growth.

The situation is even more worrying in the United Kingdom because some of the New Towns (e.g., Telford) are currently threatened by spiral decline from severe social problems. A 2002 report by the Transport, Local Government and the Regions Committee indicated that successive governments overlooked the problems related to New Towns, and, therefore, put these municipalities at risk of becoming "expensive liabilities"[34]. It is clear that the state's disengagement has resulted in difficulties to finance quality public services. To enable regeneration public aid is a necessity (e.g., through English Partnerships, the national regeneration agency), but this help is itself insufficient and must be boosted by additional funding from the private sector.

Little time, little adaptation

It is the right time to raise the question about the provision of public services since New Towns were created within a short timeframe – 30 years being the average development duration. Whereas "ordinary" medium-sized towns have had decades or even centuries to set up a wide range of public facilities, New Towns have developed them rapidly in order to become living places as quickly as possible. As urban entities designed from scratch, they did not benefit from long timeframes and prolonged social uses of space in finding the best, most appropriate solutions.

In addition, the fact that New Towns were planned according to a strict approach offers little opportunity and space to adapt rapidly to the latest changes in public needs[35]. Finally, national financing systems for education and health do not meet the requirements of fast growth. Indeed, money often becomes available after long delays whereas the development of New Towns necessitated readily available financial

---

33 Social and Cultural Planning Office of the Netherlands, *op. cit.*, p. 7.

34 See: http://news.bbc.co.uk/2/hi/uk_news/2154822.stm.

35 ENTP, *The top 8 challenges for urban regeneration in New Towns*, 2005.

means. Time has therefore meant a gap between rapidly increasing and evolving needs and a slow capacity to react.

Ageing infrastructure

The problem is not limited to the lack in infrastructures; sometimes those infrastructures exist, but they need to be restored to render them more useful and attractive for residents. This need is particularly acute in town centres, which not only provide significant employment opportunities for the residents, but also underpin the inward and outward perceptions of the town. As stated by Val Maubuée (France)[36], most of the urbanization process took place within a short period of time (in that case from 1975 to 1990), which means that the different districts reached maturity almost at the same period. With the massive arrival of socio-urban problems, a major adaptation of public equipment to current needs is called for in New Towns. Local authorities, however, have to make choices as to which buildings or districts to renovate first, provided that they have the financial means to begin with.

School in a French New Town[37], 2008

2.4. Distance and Transportation

The problem of transportation was highlighted earlier, but here we focus more specifically on the issue since it is now at the heart of many local, national and European policies concerning urban development.

36 See EGENIUS URBACT project http://www.newtowns.net.

37 Picture Edoardo Guglielmetti, ENTP, 2008.

New Towns are particularly concerned since they have experienced a period of rapid growth. In this respect, transport provision was important in connecting short distances to the major city's centre. However, the low densities of the new urban areas did not make it easy to build efficient infrastructure, and infrastructure planning has not caught up with urban development[38].

The first challenge New Towns face is that of traffic problems due to the over-reliance on commuting. Indeed, New Towns were conceived according to a polycentric perspective, as growth or job centres outside major cities where people could live. However, as will be illustrated later in the section on economic development, fewer jobs than expected were created in the New Towns. People living in these municipalities still use cars to commute to work in major cities, as in the case of the French New Towns. Jobs may be created in other New Towns, which are closer to the mother city or are part of a better transportation system, but this does not solve the problems associated with commuting (e.g., pollution and traffic jams). This communiting phenomenon is reinforced by the fact that New Towns are often very well connected to their "mother cities" at the expense of local transportation systems.

Although New Towns generally enjoy very good access and connections to their mother cities, local transportation stands in need of substantial improvements with regard to facilities, frequency and coverage of services[39]. This feature has a huge impact on other public services; while there may be a wide range of services on offer, difficulties with access will keep their use relatively low.

Such is the case in Harlow (United Kingdom), whose town centre provides many of the shopping and leisure activities of other towns. Nevertheless, the town's physical layout allows for only poor connectivity across the town centre. This is why several New Towns seek to improve their public transportation system and promote their use (e.g. Milton Keynes).

An important contribution to this book on the transportation issue has been provided by a study produced by the ENTP – European New Towns Platform – in the framework of the NEWTASC[40] project. This contextual study emphasized the role of several factors. New Towns

---

38 Gaborit P., "Reinventing the New Town mindset", in *Nova terra*, Delft, 2005.

39 Gaborit P. & Lamas Sanchez P., *Contextual analysis of six New Towns*, ENTP, Brussels, 2005. See: http://www.newtowns.net.

40 See the results of the INTERREG IIIB project "New Towns as Sustainable communities". Led by Torfaen, this project brings together the New Towns of Torfaen, Basildon, Harlow, and Stevenage in the UK, Marne la Vallée and Sénart in France and Zoetermeer in the Netherlands. This contextual analysis was made by the ENTP, on the basis of the information given by cities.

have very good access and connections to their mother cities (e.g., London, Paris, Rotterdam, La Hague). However, in terms of local transportation the connections and services have to be enhanced. Transportation strategies are aimed at curbing the use of private transportation and encouraging use of public modes. These initiatives are born out of concerns about the growth of local traffic – exemplified by the case of several UK New Towns involved in this project.

In Cwmbran, for instance, there is no real alternative to cars, limiting inhabitants' opportunities. This situation is further explained in the NEWTASC study:

> Many inhabitants in the Borough have become largely reliant on accessibility by car. As a consequence of this, there is social exclusion of some sectors of the population, especially in the south of Cwmbran. As noted earlier, the city has launched a project to improve public transportation. There is no pedestrian route between the bus and railway station and the connections between buses and trains are poor. The bus information and the frequency for train services during certain times of the day are also underprivileged. This lack of facilities will be taken into account by some projects with special attention to people with disabilities.

The town of Basildon (UK) also has excellent links with London, but it needs to improve the public transportation system between the industrial and residential areas. As underlined by the study, "the system is primarily based on east-west movement and there is a need to improve north-south links". The town is currently examining this issue further…

On the contrary, the town of Stevenage seems to be "well served by a comprehensive transport network, including rail, bus, roads and an extensive cycling path network. The town centre regeneration proposals seek to improve sustainable transport".

French New Towns, especially around Paris, are dealing with a similar situation. For instance, the city of Val Maubuée has a very good accessibility rate and is very well-connected to the outside world, but, as explained by the NEWTASC study, the town has a problem with growing local traffic. The authorities pretend to change the mind of people which usually is linked to the use of cars". The town of Sénart, 30 minutes from Paris by public transportation, finds itself in the same situation. Indeed, according to the study, while access to the city is excellent, Sénart is still trying to promote the use of sustainable transportation to fight against traffic.

Political competence, i.e., its distribution between different public authorities (concerned with transportation), is another important topic linked to the transportation issue. This means that most of the time New Towns must cooperate with other government bodies and agencies to deal with such a sensitive issue. As pointed out by the NEWTASC study in the case of Zoetermeer (the Netherlands):

> [The town] wants fast, reliable connections to the large cities in the Randstad. The city wants to be easily accessible both by road and by rail. A large section of the Randstad must be within 30 minutes to and from Zoetermeer. This ambition can only be realized in cooperation with other government bodies.

Prevailing competence issues can present barriers to creating better transport infrastructures and internal connections in New Towns.

Of course, it is important to take into account that, in light of their rapid establishment and creation, New Towns provided low costs infrastructure and transport system solutions (e.g., cycling paths are also developing).

2.5. The Rural-Urban Dimension and Landscape Management and Preservation

New Towns are specific urban areas characterised (among other features) by polycentricism and the dilution of spatial limits (i.e., between centre and periphery, natural and built areas). Often criticized as peri-urban, they also integrate another approach reflecting changing interpretations of the rural-urban dimension. It is recognized that nowadays borders vanish between towns and their peripheries. Thus, it is possible for individuals to feel "urban" and to live outside the cities. This is often implied in the creation of metropolises and agglomerations (Dunkerque, Lille, Valencienne or Aix Marseille Toulon in France, the London metropolis, or the Pearl River Delta in China). One consequence is that scales become larger for planners. Contrary to the "eco-areas" movement, the focus of which is on small scales, the creation of large agglomerations presents the "global" counterpart. But peri-urban areas also raise questions about the relationship between town and nature; more and more green spaces and corridors are integrated within the cities, whereas nature is more and more urbanized by the provision of better amenities.

As stated in a report published by UN-HABITAT: "It is now widely recognized that there exists an economic, social and environmental interdependence between urban and rural areas and a need for balanced and mutually supportive approach to development of the two areas"[41]. European, national, but also regional and local authorities progressively

[41] United Nations Human Settlements Programme (UN-HABITAT), *Promoting the Positive Rural-Urban Linkages Approach to Sustainable Development and Employment Creation: the Role of UN-HABITAT*, Monitoring and Reasearch Division, 2003.

take into account this urban-rural dimension and tend to develop potential urban-rural synergies in the areas of transport, economic activities, tourism, housing and so on. Another issue at stake is land use; urban growth and peri-urbanization often occur at the expense of farmable land and of the preservation of local biodiversity. Thus, taking into account the urban-rural relationship is a crucial condition for achieving sustainable urban development.

That is why the EU "Territorial Agenda" (adopted in 2007) seeks to enhance the "development of a balanced, polycentric urban system and a new urban-rural relationship"[42]. It is also in line with the Lisbon Treaty's renewed "territorial cohesion" objective.

New Towns and suburbs integrate within these shifting rural-urban borders. They were built on the periphery of major cities, often on farmable land. Some New Towns were formed by consolidating several rural burgs (e.g., French New Towns of Villeneuve d'Ascq or Sénart) and still enjoy a considerable rural patrimony (in terms of buildings or landscapes). As such, they must deal with land management and preservation. Before turning to a discussion about this dimension, however, it is important to note that New Towns could reinforce social cohesion and sense of community by promoting their rural past and traditions. For instance, Villeneuve d'Ascq regularly organises traditional Flemish dance activities and markets for local products.

Landscape management and preservation in New Towns have become a quintessential concern. As noted earlier, New Towns are perceived (at least initially) as the heirs of E. Howard's "garden cities". Landscapes, parks and green spaces constitute an important part of these municipalities. Moreover, they are often linked to their "mother cities" by a complex network of urban and rural areas, forests and farmable areas. Therefore, New Towns pay particular attention to the management and preservation of this complex natural heritage in order to strike a balance between nature and human activities. It further presents an opportunity for New Towns to become genuine "pilot towns".

Several reasons may help to explain why it is necessary for New Towns to focus on natural heritage management:

– Natural heritage, an essential part of New Towns: European New Towns are often responsible for managing and preserving their natural heritage – often composed of very diverse elements (i.e., landscapes, farmable lands, forests, and other green spaces). For

[42] European Union, *Territorial Agenda of the European Union – Towards a More Competitive and Sustainable Europe of Diverse Regions*, Informal Ministerial Meeting on urban Development and Territorial Cohesion, Leipzig, 2007.

example, green and blue spaces constitute more than 40% of the French New Town of Marne-la-Vallée Val Maubuée.

- This heritage is a major factor in the economic and social life of New Towns.
- Importance of green spaces: Green spaces contribute to inhabitants' well-being, providing opportunities for leisure activities. Moreover, green spaces provide "ecosystem services" which are increasingly recognized by the international community as fundamental elements for the survival of present and future generations[43].
- Vehicle for citizen participation and social cohesion: Landscape management and preservation also present an opportunity to foster citizen participation and social cohesion. Indeed, as pointed out by the architect Lucien Kroll, if local authorities want to manage and preserve their natural heritage in a sustainable way and create a genuinely "natural city" (rather than inviting nature into the city), they must involve local inhabitants. This involves understanding their desires, needs and perceptions of everyday life in the city. Further, as New Towns' inhabitants are the direct users of landscapes[44], they must be involved from the beginning in landscapes management processes. This approach both increases awareness of the value of living landscapes and strengthens social cohesion by involving local populations.
- Image and identity: Last but not least, sustainable landscape management is a major part of city image and identity. This aspect will be examined later in the book, but, as we will see, identity is becoming a major issue for cities, – "old" medium-sized towns as well as for New Towns – in an increasingly globalised economy to attract investments, companies, and high-income inhabitants. Thanks to sustainable and well-designed natural heritage management processes, New Towns have a unique opportunity to polish their image.

However, this issue is often misunderstood and inadequately considered by researchers and local authorities who tend to focus on a narrow definition of landscapes (i.e., reducing them to urban green spaces). The European Landscape Convention Treaty, adopted by the Council of Europe in 2000, is the first European text to offer a comprehensive

43 The ecosystem services were defined and popularized in 2004 by the United Nations. See United Nations, "Millennium Ecosystem Assessment (MEA)", in *Ecosystems and Human Well-Being: Synthesis*, 2005, Island Press, Washington.

44 Kroll L., "Quelle place pour la nature en ville?", in *La ville autrement*, Alternatives Economiques, Paris, 2009.

vision factoring in the concept of natural patrimony, but as yet, there is no harmonised European instrument for the sustainable management and preservation of such heritage. In France for instance, local authorities can develop what is called a SCOP (Schéma de Cohérence et d'Orientation Paysagère: Master plan for landscapes). Local authorities can also exchange good practices and learn from each other through European projects, such as the EUROSCAPES project, among others[45].

2.6. Housing

"A large part of the present development of European cities is vehicled by the expansion of residential areas"[46]. This was especially the case in New Towns of North Western Europe that were built in response to the population growth's increasing pressure on housing at the end of World War II and until the end of the 1980s.

2.6.1. Housing Still on the Growth?

The original aim of New Towns was to respond quickly to intense population growth. Therefore, the top priority of national governments was to provide homes at moderate prices to accommodate growing housing needs. In France, the Netherlands and the United Kingdom the development of New Towns stemmed from an interventionist approach on the part of the state with clearly defined growth targets set beforehand. As a consequence, this original goal led to a dominance of housing, which today largely accounts for the "dormitory town" image. New Towns have mostly managed to accommodate increasing needs.

However, the demographic trends that were expected at the creation of those towns proved to be exaggeratedly high, requiring several adjustments over the years. This was particularly the case in France, but also in the Netherlands, where 20% of growth centres were to be compromised of new buildings. This target was never achieved, however, and the maximum level was 15% by the early 1980s[47].

---

45 See the EUROSCAPES project website or the ENTP website: http://www.newtowns.net.

46 Urbact II Working group Hopus, Housing praxis for Urban sustainability, *Good Green, Safe, Affordable Housing*, 2009, p. 15.

47 Social and Cultural Planning Office of the Netherlands, *op. cit.*, p. 4.

Case study: the Vinex Plan (the Netherlands)[48]

– Every ten years, the Ministry responsible for development issues a report setting the framework for the development of the country as a whole. Following the fourth report, the Vinex Plan (1995-2005) committed the country to housing growth. The aim was to respond to demands for more home ownership and larger homes.
– Thanks to the Vinex Plan, the housing stock increased by 7.5%, i.e., a total of 455,000 new homes among which 285,000 are extensions of existing towns and cities. Over half that growth was concentrated in the main conurbation, namely the Randstad. The plan provided an overall amount of 4 billion Euros, of which 70-80% was allocated for infrastructure. In fact, the amount per home was quite limited and required local authorities to invest a lot of money.
– The Vinex Plan largely explains why housing construction in the 1990s and the beginning of the century remain considerable in New Towns. In fact, the scaling down of growth centres policy has not meant the end of state financial aid. However, this aid does come with strings attached, as such funding is specifically designed to cover the construction of new housing.

Recent demographic changes, however, have called into question the focus on large scale housing development. The role of New Towns as suppliers of new housing is coming to an end, or at least slowing down, reflecting local situations and needs.

Some New Towns no longer have space available to launch major housing projects and must increase density in order to cater to more residents. Others are no longer considered attractive places to live and are acutely in need of urban renewal. The main challenge faced by New Towns is the adaptation of their urban structure and what they have to offer (in terms of housing) to current changes. As in most European suburbs, housing was designed for traditional families, but the increase in divorces and single-parent families has triggered new housing needs.

The problem is that national financial aid is scarce or completely cut off (with the gradual normalization of the status of New Towns), meaning that municipalities increasingly bear the costs of new housing programs. Jean-Pierre Marchetti, Director of Services for the French New Town of Val Maubuée, reports: "the national government expects us to build 60,000 to 70,000 new housing units in the forthcoming years, but we are lacking the financial means to reach this target".

48 *Learning from Dutch new towns and suburbs: report of the Harlow Renaissance study tour*, Harlow Renaissance Ltd, 2007.

A few New Towns, on the other hand, seem to stand out in this respect because they still experience strong growth. Such is the case for Almere, which has grown from scratch to a community of more than 160,000 people within 25 years. In 2001, the Dutch government proposed a further expansion of the town, up to 400,000 inhabitants by 2030. Almere had accepted the idea of fast and strong growth, provided that the central government chip in to make it an economically, spatially and socially balanced town. The most interesting national policy in terms of housing is to be found in the United Kingdom. Indeed, the Sustainable Communities Plan (ODPM, 2003) has identified four growth areas[49] where 240,000 new homes, over and above previous regional plan targets, shall be delivered to reduce the shortage of housing in the South of England[50]. Yet, in an entirely different context, the British New Towns programme is the largest building strategy – in both scale and ambition – the UK has experienced. Several New Towns[51] (among them, Milton Keynes, Harlow and Stevenage) are directly concerned by the plan, a major lever in boosting their development. The British example shows that New Towns continue to be viewed as a solution to housing shortages and poor housing conditions, even though this new policy was not designed specifically for them. On the other hand, after years of neglect, their potential is nevertheless acknowledged.

Comparison of New Towns and Growth Area policies[52]

New Towns Programme	**Sustainable Communities Plan**
Similarities:	
– Objectives to increase housing supply, relieve housing pressures in overcrowded areas and provide good quality housing in specific locations	– Objective to achieve a "step by change" in housing supply and accommodate economic growth
-A national programme, although not fully joined-up with regional policy, part of wider housing growth targets	– Part of a national strategy with distinct regional objectives
Differences:	
	-Emphasis on brownfield development
-Developed mostly on Greenfield sites	-Medium to high density
-Low density	-Integrated with existing communities

49 The growth areas are the Thames Gateway, Ashford, Milton Keynes-South Midlands and the London-Stansted-Cambridge corridor.

50 http://www.communities.gov.uk/housing/housingsupply/growthareas/

51 http://www.communities.gov.uk/documents/housing/pdf/898634.pdf/

52 This table was published in Bennett J., *From New Towns to Growth Areas: learning from the past*, Institute for public policy research, London, 2005, p. 6.

Intended to be self contained communities -Initial focus on decentralization of people and jobs away from London -Public sector led development -Total target population increase over 1,500,000	-Accommodating household and economic growth across three regions, alongside significant growth within London -Public-private partnership approach to development – Target increase in the number of homes over 200,000 by 2016

2.6.2. Urban Regeneration Needs

Urban regeneration in New Towns is an increasingly important issue. Even though it has been tackled by concrete actions rather recently, this impending challenge was already being pointed out in many reports and articles in the 1970s. For instance, by 1976 British New Towns were said to be showing their first "wrinkles"[53]. As European New Towns were conceived and developed mostly from the 1950s to the 1970s, they are typically thirty to forty years old by now. Therefore, they need more than simple cosmetic maintenance, especially with regard to housing.

In any case, both those concerned with large-scale building programmes and those that have stopped growing share the challenge of regenerating the oldest areas in their carefully planned urban environment. The challenge is all the more important as New Towns are experiencing a normalization of their special status, which in concrete terms means managing without state support. Gunilla Edwards, economic officer of the British New Town of Basildon[54], explains further:

> We are selling some land as government, but with the financial crisis we have few buyers. The challenge is also to keep private developers involved in developing the town. We also have a problem of empty flats; people sometimes buy property only to make investments and then let them empty.

The ageing housing stock has resulted in several problems for New Towns:

– The poor physical quality of housing units[55] needs to be emphasized, especially in regard to early post-war New Towns and neighbourhoods. The buildings' initial design and the materials

53 Champenois M., "Les rides des *new towns*", in *Le Monde*, 10 February 1976.

54 This quotation comes from an interview realised in 2009,

55 Czischke D. (ed.), *Urban regeneration in Europe: the place of social housing in integrated urban policies*, CECODHAS, 2009, p. 27-28.

used in their construction have been criticized for their poor quality and for being out of date.

- This lack of quality could have resulted from the approach to internal design and structure (e.g., sanitation equipment, inadequate central heating, etc.) or from the approach to urban design itself (e.g., high density, inaccessibility of the estate, etc.). Moreover, physical decay is highly visible to the residents of the estates, creating a feeling of discomfort and discontent.
- Due to the poor quality of housing, some areas are confronted with a de-cohabitation and a de-densification of the housing stock, as well as a correlated population decline. For instance, the population of Vällingby has decreased from 29,000 to 13,000 inhabitants within two decades[56]. In parallel, New Towns have started to experience high concentrations of low-income households with low levels of education. As discussed earlier, the physical planning of New Towns did not necessarily translate to socially balanced communities. Although they tended to attract middle-class households at the onset, they now host a high level of social housing and are plagued with pockets of deprivation. This phenomenon can be attributed to the outward migration of those who can afford to leave followed by the inward migration of those who have fewer financial means and possibilities (e.g., low-income households with a lower level of education)[57]. New Towns have seen in recent years the arrival of many foreign migrants with large families, which has also influenced the socio-spatial distribution of the population. Although they managed to cater to population growth, their success in terms of development quality is questionable.

Poverty indicators are indeed a matter of concern, especially for British New Towns. For instance, in Harlow, the third most deprived district in the region, (as reported in 2007 by the NEWTASC and SET4U projects studies), "employment is slightly lower than the Essex average, [but] there is a significant level of social deprivation".

Residents' education or work opportunities are significantly diminished due to the isolation and decay of some areas. Therefore, many safety issues have emerged alongside social deprivation. According to Don Burrows of the Neighborhood Initiatives Foundation[58], the original

56 Baudelle G. & Ducom E., "Renouveler les villes nouvelles?", in *Revue Urbanisme*, September-October 2008.

57 Czischke D. (ed.), *op. cit.*, p. 30.

58 The Neighbourhood Initiatives Foundation is a Telford-based charity which aims to help improve local communities.

design of some neighborhoods is the source of many of the problems: "High density large estates with poor amenities, mainly designed for the car not the pedestrian are turning into breeding grounds for petty crime and drug abuse"[59].

This problem is particularly acute in regard to youngsters, since youth unemployment and school drop outs often generate anti-social behavior. Fostering social cohesion is all the more important in those areas because it could help cultivate a greater sense of security and help prevent physical deterioration of neighborhoods.

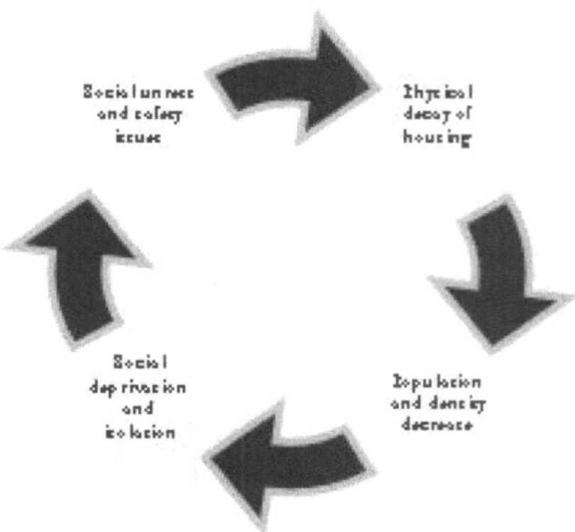

Regeneration of housing is therefore required in order to prevent such spiralling decline. However, this is no easy task: how do we choose between demolition and refurbishment, for instance? Although an expensive and sensitive process, demolition is a rather efficient tool to speed up regeneration. According to accounts by French New Towns, large housing settlements in French suburbs are demolished and rebuilt under the National Agency for Urban Renovation (ANRU)'s supervision.

Since the New Towns were established by and belong to public authorities decision-making is easier. Refurbishment, referring to physical measures taken to improve dwellings' structure and surroundings, is a simpler process[60]. Refurbishment of housing is necessary. This is particularly so for social housing, which accounts for a considerable percentage of housing in New Towns (e.g., 30% in Milton-Keynes) and

[59] See the article: http://news.bbc.co.uk/2/hi/uk_news/2154822.stm/

[60] Czischke D. (ed.), *Urban regeneration in Europe: the place of social housing in integrated urban policies*, CECODHAS, 2009, p. 29.

which typically suffers from significantly greater degradation than private residential units in high income areas.

However, several New Towns[61] have pointed out the difficulty of refurbishing housing under shared ownership or co-property arrangements earned by low-income groups. Indeed, it is difficult for local authorities to target this group, as in the case of Val Maubuée (France). The town's Director of Services stated in 2009:

> Val Maubuée has no large housing estates. The main problem we have to face is the refurbishment of the housing stock. In French suburbs, the process is easy; here, the important share of co-ownerships in decay is another matter that raises important management issues. Indeed, it is not possible to force people to renovate or leave.

This problem is particularly serious in Central and Eastern European countries, where most of the housing stock is based on co-ownership arrangements. In this situation, where urban regeneration is increasingly needed, how can we foster awareness-raising and public action?

Another crucial matter linked to regeneration is the quality of the improvement, i.e., the sustainability of operations and the concrete benefits derived from them. Currently, many initiatives are aimed at improving the energy efficiency of the buildings. Local authorities in the Ile-de-France region (France) have decided to invest heavily in this area. For example, Saint-Quentin-en-Yvelines has established a local energy agency operating at the regional level and developing pioneer initiatives targeted at inhabitants, while Val Maubuée is involved in a pioneer initiative in geothermal energy, potentially enabling the town to set up a local fund to help households make energy audits and energy savings work. The Swedish government in 2004 adopted a strategy for the energy performance of buildings[62]. The New Town of Katrineholm directed a study called Energy efficiency measures in Swedish detached houses: perceptions and attitudes of homeowners[63] and subsequently decided to lower the town's energy demands for both environmental and economic reasons. 13,770 insulation panels were installed on all the town's schools and child care facilities; returns from investments are expected within 8 to 12 years. Milton Keynes (UK) has developed a pioneer initiative called the Carbon Offset Fund. This fund was created by taking a percentage out of newly created housing estates; it will co-finance actions for the refurbishment of older housing areas.

---

61 This is the case especially in the French new towns.

62 See in: http://www.sou.gov.se/mvb/pdf/Energieffektiv%20bebyggelse%20engelsk%20version.pdf.

63 See in: http://www.energimyndigheten.se/Global/Filer%20%20Forskning/AES/Forskningsarena%20Katrineholm%2025-26%20aug/Leif_Gustavsson.pdf.

2.6.3. The Paradoxes of Diversity

Long considered places where people can meet and mingle cities are now viewed as vectors of social segregation and exclusion. Diversity has therefore become a recurrent and dominant leitmotiv in spatial planning discourse; in other words, achieving diversity of functions, as well as mixed social backgrounds, age groups, cultural backgrounds and nationalities appears to be the right thing to do. Most planners and politicians have definitely abandoned the segregated approach, replacing it with concepts of "conciliation" and "togetherness". Yet, in this new framework, New Towns and their inhabitants still have a long way to go. As underlined by Jukka Kullberg, Head of Planning of the Finnish New Town of Vantaa, in a 2006 interview: "Social diversity is not easy to achieve because a significant number of inhabitants want to live where this social diversity does not exist; they want to have their own house in an area of houses only".

As for large housing settlements, there is obviously no comparison with early post-war neighbourhoods known as "banlieues" (suburbs) in France.

Those areas consisted almost entirely of social housing, lacking their own employment opportunities, services and facilities; as a consequence, they rapidly became "dormitory towns" plagued with social problems[64]. But in addition to lessons being taken from this experience, New Towns were endowed with their own individual identity. Although medium and large housing estates are relatively common, they often appear alongside rental housing and owner-occupied homes. This is in line with the concept of polycentricism elaborated earlier – a core idea behind New Town urban design. These new poles attracted new inhabitants that we call "neo-urban". There are also different centres for consumption activities (e.g., shopping malls), work (e.g., techno poles) and entertainment (e.g., cinemas, green parks, leisure centres and sport stadiums).

Polycentricism had paradoxical impacts on segregation (with the creation of ethnic areas or ghettos), but at the same time encouraged a more social mix[65]. Indeed, large housing estates and individual housing units are seldom located in the same place, which has led to differentiated spaces with one social group and spatial monotony dominating each area. During periods of fast growth, thousands of inhabitants had to be accommodated in a very short timespan (e.g., 75,000 inhabitants

---

64 Molster A., "New Towns revisited", in *'SCAPE*, 2006.

65 Gaborit P., "Development of New Towns and peri-urbanized areas", in *International conference on New Towns Report*, Ministry of Housing and Urban Development in Iran, Tehran, 2005, p. 282.

within 30 years in Val Maubuée), meaning that entire districts were created with exactly the same type of buildings. This forced accommodation entailed an over-reliance on housing estates, which eventually resulted in a negative image difficult to shake. Even though densely built neighborhoods conjure up negative images in people's minds, the recent shift in urban planning, which lauds the advantages of a "compact city", raises new questions. Are large housing settlements fundamentally inadequate and prone to social problems, or did the current situation stem from a bad approach at the onset? Can high density be achieved without jeopardizing social cohesion and diversity?

With or without large housing settlements, New Towns are having difficulty attracting other social groups (i.e., more middle-and upper-class households), preventing a more diverse social make up. In general, the proportion of rental housing to privately owned units is high.

This is particularly so in the French New Towns of the Ile-de-France region, where the high proportion of young people (owing to the presence of universities) has significantly reoriented the housing demand towards more affordable units. Dutch New Towns stress the importance of the luxurious housing and rental market, which could potentially replace a proportion of the flats in the social housing rental sector. Such housing is called for and actually increasing, but still represents a very small proportion of available housing[66]. Providing a satisfactory housing option for the middle-class is a cross-sectoral task that requires adjustment of the housing stock, as well as the services and amenities used by the population.

While some New Towns think it is good policy to mix people from different social backgrounds in the same neighborhood or the same building complex, others tend to reject this idea. They argue that it is up to the inhabitants to make social diversity work. Echoing critics[67], they point out that discordant relations cannot be subdued by the simple fact that different kinds of population are liaised in a supposedly pacified atmosphere.

Diversity is all the more acute as the price of housing has become a major problem for many New Towns. Their proximity to big cities has once again exposed negative side effects. Indeed, even though prices remain relatively low compared to major cities, there has been a steady increase threatening already vulnerable households. This is particularly the case for British New Towns. Harlow (United Kingdom) has experi-

---

66 Social and Cultural Planning Office of the Netherlands, *op. cit.*, p. 4.

67 See the article in: http://www.laviedesidees.fr/Pour-une-approche-critique-de-la.html/

enced massive increases in housing prices against people's incomes[68]. It is one of the most deprived towns in the east of England, and current housing prices are too high and unaffordable for many households who would otherwise wish to purchase their own home. In Stevenage the situation is even worse.

According to a local housing market survey carried out in 2000, a gross income of more than £26,000 is necessary to buy a family home at the lower end of the market, whereas the average household income is £13,405. As a consequence, most of the families cannot afford to rent privately or buy property in the Borough[69].

In addition to problems of regeneration, social and economic difficulties can create a homelessness phenomenon. This is why some New Towns have developed supporting initiatives like the foyer in Harlow.

Case study: to accommodate low income groups:
Harlow foyer (United Kingdom)

– Harlow Foyer, also known as Occasio House, opened in 2001 and **provides local services to residents and other young people living in and around Harlow**. The foyer works to move young people towards independence and operates an accommodation programme of 116 units. The maximum length of stay is 2 years.

– The foyer has staff available everyday to support residents. Keyworkers help residents identify the issues they need to address, plan approaches for tackling certain issues and advocate and support residents in accessing specialist services whenever these are required (e.g. lifeskills training sessions, careers advice, young parents support).

– Harlow Foyer is **a major actor in the integration of youngsters in their local environment**. It is an efficient tool for preventing the social marginalization of young people and the security issues linked to this phenomenon.

2.7. Economic Development

It is commonly agreed that a municipality's economic development is critical for attracting inhabitants and generating wealth, especially in a globalized world where competitiveness has become a paradigm. Strong economic performance relies on the capacity to attract business,

---

68 Gaborit P., "Contextual analysis of six New Towns", NEWTASC INTERREG III B project http://www.newtowns.net.

69 *Ibidem.*

but also to generate innovation and entrepreneurship and to ensure a good level of connectivity.

2.7.1. Problems Encountered by New Towns

The issue of economic development is particularly sensitive for New Towns, which were primarily created to absorb the high population growth of the post-war period. Consequently, priority was given to the provision of housing and public services at the expense of economic activities.

The fact that New Towns benefit from good connections to major cities could be regarded as an asset, since it implies that they are potentially good locations for businesses to relocate[70]. And indeed one of the landmark measures to attract businesses was the creation of large business parks, i.e., spaces specially designated for business activity, with tax facilities and other services. Although many New Towns have launched active policies to stimulate their economic growth and attract businesses, they have experienced difficulties doing so. Economic development has translated into good results in towns such as Milton-Keynes, where the workers to residents ratio is highly positive. However, the lack of diversified economic activity, which is otherwise a boon for local employment opportunities, has negatively impacted the image of New Towns:

- In the towns where the workers to inhabitants ratio is negative the local authorities experience a lack of revenue, especially in countries where business taxes were local authorities' main source of income (e.g. France). In such cases, attracting businesses is not only a matter of image, but foremost a matter of financial existence.
- A majority of New Towns residents have to work outside of their town and commute on a daily basis. This poses an obstacle to developing economic competitiveness, but also to fostering an environment of well being for residents.
- This phenomenon of commuting exacerbates car traffic and car-related issues (e.g., traffic jams, pollution). Studies carried out in Ile-de-France have shown that the longer the distance between home and work for New Towns' residents, the more they use their car[71]. Even residents who work in their town tend to use their car because New Towns often sprawl across a wide terri-

[70] Gaborit P. & Lamas Sanchez P., Contextual analysis of six New Towns, ENTP Brussels, 2006. See: http://www.newtowns.net.

[71] Beaufils S., *Les temps de transport des actifs de villes nouvelles proches de ceux de grande couronne*, IAURIF, 2005.

tory, yet their public transportation services are usually insufficient. Distances between places of residence, schools, retail areas and leisure centres make it impossible for inhabitants to rely solely on public transportation. What's more, the absence of cycling paths in some countries does not help remedy the situation.

- Finally, the "business park" economic development model has led to image problems. New Towns seem like "dormitory towns" where nothing happens during the day, which is likely to scare off businesses and investors.

2.7.2. Innovative Strategies

New Towns have well understood that they need a multi-functional approach to economic development in order to remain attractive to residents and improve their overall image. However, they are facing several challenges in achieving this goal[72].

Challenge 1: Attracting more (innovative) businesses

New Towns often suffer from their proximity to large cities; the dynamism which major cities are typically associated with pushes businesses to set up shop there, rather than in New Towns. The NEWTASC and SET4U project studies reported the following with regard to Harlow (United Kingdom) in 2007:

> There are 1,570 businesses in Harlow; this is a remarkably small number given the number of economically active residents. Harlow needs a substantial increase in its business growth rate and a stronger SME sector.

The likely result is a lack of local jobs compared to the number of active inhabitants, effectively creating an imbalance between housing and jobs.

Another phenomenon observed in some New Towns is the challenge of filling spaces in business parks, which are comparatively unappealing when measured against their competitors in nearby major cities. The Dutch New Town of Capelle aan den IJssel presents such an example, as reported by the SET4U project's proposal in 2004:

> In comparison to the other surrounding cities in the region, Capelle has the highest number of empty office buildings. This is a severe problem and it implies a large job loss.

In any case, New Towns have launched strategies to diversify their economies and endeavour to attract innovative businesses, which constitute a significant source of revenue and contribute to a positive image.

---

72 The quotations come from the application form of the project SET4U INTERREG IV B project, 2007.

These strategies often focus on innovate technologies, but it is difficult to draw comparisons; while some towns attract more services and new high tech companies (e.g., Nokia in Espoo, Finland), others, like Sénart (France), tend to attract logistics and transport companies because they offer more space.[73]

Case study: Advancity competitive cluster in Val Maubée (France)

– When it comes to sustainable development, Marne-la-Vallée/Val Maubée seems like one of the most advanced innovation centres in France. The Advancity cluster is aimed at **achieving sustainability in cities through 3 interlinked themes**: city and city planning, housing and construction, mobility and transport. The cluster is composed of a network of more than 100 members, including 10 global companies (e.g., EDF, Veolia, Suez, etc.) and renowned universities and research centres (e.g., Cité Descartes, École Nationale des Ponts et Chaussées, etc.).
– The cluster was created in 2006 with the **following aims and outcomes**:
- Achieve a leading national and international position in economic activities related to sustainability
- Contribute to the fight against climate change: indeed, the cluster's targets (i.e. building and transport) account for 2/3 of the carbon emissions

73 Pictures by Eric Morency.

• Strengthen positive image and identity of the territory
– The cluster has approved more than 60 projects for a total amount of 130 millions euros, which makes it **the 5th most successful French competitive cluster in terms of financed projects.**
– **What are the replication possibilities?** The implementation of a competitive cluster provides the opportunity for a New Town to become nationally or internationally renowned in a high added-value field. It is thus an efficient tool for creating qualified jobs and for attracting both businesses and inhabitants. Nevertheless, clusters need to rely on a solid basis of disposable resources, i.e., the proximity of already competitive (or promising) businesses and research centres. It is particularly evident for Advancity, the success of which is based on many renowned universities and research centres (e.g., Cité Descartes). Moreover, the cluster is part of a comprehensive national policy that enables businesses to benefit from funds and fiscal exonerations. The state is an active member, providing vital financial resources.

Challenge 2: Deal with the lack of jobs and employability

Some European cities experience a paradox where job opportunities do not necessarily meet local demand. Most New Towns are confronted with this challenge because there is typically a huge gap between the needs of business and the skills of the local labour supply. Despite increases in the number of jobs available, it is hard for New Towns to find and retain highly skilled workers. Indeed, their high number of social housing demonstrates that a considerable proportion of the population is low skilled.

The British New Town of Stevenage is a case in point. The SET4U project's proposal[74] stated in 2007:

> New business registrations are significantly lower than the regional and Great Britain averages. Businesses struggle to recruit and retain highly skilled workers and it is predicted that the town's economy will barely grow over the next 20 years.

Therefore, the local population's employability must be improved, but this is not simply a matter of increasing job numbers. Rather, the point is to ensure that available jobs benefit the local population, enabling an ever growing number of local individuals to access the labour market. New Towns must not only have the capability to attract highly skilled workers, but to provide the local population with jobs matching people's education levels and skills. The current frenzy to develop competitive clusters, for instance, must consider retail and shopping needs, which are likely to entail a significant amount of local job crea-

74 EU project's proposal about creation of innovative economic strategy that was finally not approved (2007-2008).

tion. In any case, improving employability should enable New Towns to reduce outward commuting, which sometimes reaches a high level, as in the case of the French New Town of Sénart, where there is a deficit of employment despite comparatively weak unemployment figures (both regionally and nationally). As explained in the SET4U project's proposal in 2007, "75% of the population is commuting to work outside Sénart for Paris or other places".

Challenge 3: Overcoming the ICT[75] gap

Access to information and communication technologies is recognized as a key ingredient for cultivating knowledge economies. Cities and towns have to provide efficient, accessible and affordable infrastructures to meet the needs of businesses and public services. They also bear the responsibility of ensuring that no social group is excluded from access to ICT, e.g., by developing community-based access points available to all and providing support and training systems, particularly in disadvantaged areas. As host to a globally recognized ICT school, and with 20% of its population having an ICT-related job, Zoetermeer (the Netherlands) is known as the Dutch ICT city. Many other examples show that New Towns are increasingly investing in this field.

Case study: Very high speed internet access
in Saint-Quentin-en-Yvelines (France)

– In a world that relies increasingly on the opportunities and benefits afforded by the knowledge economy, internet access has become quintessential in cities' quest to demonstrate that they are indeed major economic players. This is why Saint-Quentin-en-Yvelines intends to **become a leading digital planner via an optical fibre networks project**. The network was launched in 2009 and the entire agglomeration will be equipped with optical fibre technology by 2011.

– The project is aimed at attaining the **following goals**:

- Anticipating future technological needs and reinforcing the New Town's competitiveness
- Ensuring that all inhabitants and businesses (even the most remote among them) can benefit from this service

– The project is a good example of how a strategy can materialize. The New Town has understood the potential benefits of ICT and endeavors to turn its former weaknesses (e.g., exclusion of areas, unique operator) into assets (i.e., optical fibre as a way to attract more businesses and simultaneously promote social inclusion). Thus, the project manages to **associate economic competitiveness with local development**.

– **What are the replication possibilities?** It is essential that such an initiative

75 ICT meaning Information and Communication Technology.

involves the whole territory, no matter how small some of the urban entities can be. Indeed, lack of high speed internet access should not be a reason to exclude some territories that are considered "less profitable"; rather, they should be brought into the fold. The municipality must be the driving force behind such a project, ensuring that there are no spatial (and social) inequalities in the delivery of the service.

Challenge 4: Achieving balance between competitiveness and quality of life

The initial success of New Towns was thanks to the high standards they had set (e.g., quality of life involving the presence of nature and provision of public services). Developing a competitive economy must not be done at the expense of quality of life, which is why many New Towns try to highlight their quality of life in addition to their innovation-friendly environment.

Case study: Sénart's economic situation (France)

– Between 1990 and 1994, New Towns located in the Ile-de-France region experienced a 15.3% increase in jobs; in other words, they catered to 42,300 new jobs during the period[76].
– The development of New Towns has not been uniform throughout the Ile-de-France region. Sénart has experienced great difficulties because of its wide territory, its low densities, its remoteness from Paris (30 km) and its lack of centrality[77]. Today, Sénart still suffers from a job deficit (only 6 jobs offered for every 10 residents), whereas most of the other New Towns managed to achieve a balance between jobs and housing.
– Nevertheless, Sénart had one of the highest job growth rates between 1990 and 1994, i.e., a 19.6% increase with 3,820 jobs created[78]. Furthermore, the city is diversifying its economy in order to become more services-oriented, catering to an increasing number of SMEs and international companies[79].
– This economic boom, however, does not counter a high quality of life. Sénart calls itself the "town in the countryside" and offers many opportunities to enjoy nature and local heritage. What is more, the charter of environmental quality ensures that businesses contribute to environmental protection.

---

76 Fouchier V., *La politique des villes nouvelles (1965-2000)*, programme interministériel d'histoire et d'évaluation des villes nouvelles françaises, 2000.

77 Beaufils S., 2005, *op. cit.*.

78 Fouchier V., *La politique des villes nouvelles (1965-2000)*, programme interministériel d'histoire et d'évaluation des villes nouvelles françaises, 2000.

79 See: http://www.epa-senart.fr/admin/data_/pdf_cp/CP_TertiarisationSenart.pdf.

New Towns are becoming increasingly cognizant of these new challenges, as reflected in current developed policies, which have undergone major changes over time. After a period between 1960 and 1980, when governments tried to attract firms by offering cheap factor inputs and providing subsidies or tax reductions, a second stage in the 1980s was marked by development of education and technological infrastructure in order to cultivate a knowledge base supposedly making way for a competitive response to the changing economic environment. However, the quality of physical, social and knowledge infrastructures has become the primary factor in investment and business location decisions; thus, this third phase of economic development accentuates the importance of investments in soft infrastructure, public-private partnerships and highly targeted inward investment in adding to the competitive advantages of local areas[80].

2.8. Social Exclusion

Peripheries can be a symbol of segregation and exclusion for some people, but not for others. They are the places where rural or foreign migrants have to be, away from "local society" and sometimes denied the future prospect of settling in the city centre. In general, the process is different in New Towns, which were conceived for individual families in a spirit of equality.

In the late 1970s, during the establishment of New Towns and new growth centres, the construction of large housing settlements and the separation between the middle class (individual housing units) and the lower classes (in the housing estates) reflected a period of confidence in progress. The objectives were to narrow differences and to reduce conflicts. In reality this did not turn out as planned. Indeed, the initial idea behind New Towns was to create good housing conditions for workers and the middle class. However, during recession, the perception of non-qualified workers changed and immigrants were stigmatized. Thus, housing paradigms have resulted in the relegation of the poorest people and of the ethnic minorities. Indeed, as observed by J. Hobson: "The emphasis on this type of housing (for nuclear families), can be seen as encouraging suburban conformity, through designing the built environment, so that other household forms (single person or homosexual adults sharing) are difficult and even appear deviant[81]".

If we look at poverty indicators – a rather basic way to examine marginalization – particularly in towns which have been created with the intention of improving conditions for the urban poor, New Towns

80 Introduction to the ENTP General Assembly, 2008.

81 Fouchier V., 2000, p. 13.

appear only half successful. In Milton Keynes, for instance, "[t]he harsh realities of life for those in some of the large public sector housing estates, where poverty and unemployment are endemic, also help to define to which extend those residents are marginal in the overall project[82]". In Harlow we find the same patterns: "While employment is slightly lower in Harlow, than in Essex average, there is a significant level of social deprivation, with Harlow the third most deprived district in the region"[83]. The local government's improvement program, in Stevenage states:

> There are areas of socio-economic need and relative depravation". The borough has the highest rate of teenage pregnancy in the county and the Eastern Region. There is also a high percent of single parent households. Health is another key issue for the authority, as the death rates for the age cohorts under 75 are higher in Stevenage than the rest of Hertfordshire[84].

Have social policies counterbalanced these effects in New Towns in the long run? This question is rather controversial. The Memorandum submitted by the Towns and Country Planning association in the United Kingdom states:

> It is not established that social exclusion in the New Towns is more or less extensive than in ordinary places. The association's observation is that the New Towns generally achieved their objective of accommodating all varieties of households and those most in need of a home and a job[85].

This point of view is shared by civil servants working in other New Towns.

2.9. Identity and Sense of Community

> Branding and marketing to support image and identity can be a force to bind and attract people and businesses as well as to support the cities in engineering social and economic transformation of urban areas (...) For identity to be used as an asset, it must be managed, anchored, nurtured, safeguarded and manipulated[86].

---

82 Hobson J., "New Towns, the modernist planning project and social justice: the case of Milton Keynes UK and 6th October Egypt"; University Coll, London, 1999, p. 14.

83 "British local government improvement programme: visit to Harlow District Council", 19th-23rd March 2001.

84 "British local government improvement programme: visit to Stevenage Borough Council", 13th-17th January 2003.

85 Town Country Planning Association, "The New Towns: Their problems and future", Memorandum 2002.

86 Scheffler N., Kulikauskas P., Barreiro F., "Managing urban identities: aim or tool of urban regeneration?, in *URBACT TRIBUNE*, 2009, p. 13.

As we venture into the issue "image and identity, we move away from a strictly technical treatment of urban matters and approach studies of subjective representation, of imaginaries shaping daily life and closer to the sociology of imaginaries[87]. In light of his approach to esthetism, Simmel is probably one of the sociologists best illustrating how this notion of imaginaries applies to social life in general and to the life of New Town inhabitants in particular. He recalls that social cohesion relies on an "imaginary focus", induced by conflicts, diversity of representations, attractive and repulsive forces between individuals[88]. These representations are linked to reality by a certain type of esthetism that is sometimes prevalent in a town's architecture and heritage (e.g., he mentions Florence or Venice)[89]. But these representations also impact different policies; Mongardini (2004) was indeed one of the few authors to demonstrate that fear is one main representation of society and one of the main motivations behind local policies (i.e., security, health, indicators)[90].

2.9.1. Why Is This so Fundamental?

Cultural and spatial symbols help produce identity and bestow a special atmosphere (hopefully) distinct from other urban entities. As such, the image reflected by a given town or city is not to be underestimated. It is a crucial undertaking to provide a sense of belonging for local residents and pave the way for social cohesion, all the more as businesses increasingly take into account the local social climate.

At a time of growing competition between cities and global flows of culture, tourism and business activity – in which the export and import of products, cultures, ways of doing things, etc. erase away at long-preserved identities and traditions – the issue of uniformity must be raised. Indeed, the value of cultural and spatial symbols or a place's identity is an important competitive factor. Moreover, cities constantly evolve as a consequence of competition, complementarities between town and periphery and the global economy, in addition to mass-media proliferation of cultural values and the global flows of business travelers and tourists. The result is that cities are becoming more and more uniform, thereby, loosing their local character. Furthermore, central urban areas are allocated higher priority at the expense of less spectacu-

---

87 Legros P., Monneyron F., Renard J.B., Tacussel, *Sociologie de l'imaginaire*, Armand Colin, 2006.

88 Simmel G., Conflict and the group of web affiliations, New York Press, 1955.

89 Simmel G., *La philosophie de l'imaginaire*, (Fr version), Paris, l'Arche, 2002, p. 47.

90 Mongardini Carlo, *Le Dimensioni sociali della parura*, Milan, Franco Angeli 2004 quoted by Legros, Monneyron et alter Opcit.

lar neighbouring cities like New Towns. This development indicates a relationship between local preservation and globally inspired renewal. At the same time, the identity of a place provides a feeling of belonging and meaning for local inhabitants.

Therefore, the key questions to reflect on are: what makes a town or an urban area unique? And as far as New Towns are concerned, do they portray a specific identity or do they try to be "mainstream" towns? The current trend of giving financial priority to central urban areas, like the regional agglomeration of "Grand Paris", also raises several problems in terms of spatial position and influence: what happens to less attractive and less dynamic peripheries?

However, the dynamics of identity in the context of globalisation are not always biased towards capital cities and metropolises; the globalisation trend can be used constructively (Nguyen 2009)[91]. By making the scenery, architecture, lifestyle and culture the basis for development (Acted 2009[92]), and at the same time, drawing from new trends, it is possible to see the local and global character reflect one another. In this way, it is possible to valorise local character in both the centre and in the periphery for the benefit of local and global actors.

Therefore, consideration of the spatial environment is fundamental if the identity of a place is to be created, developed and promoted (Ruggeri, 2009[93]). Highlighting local identities in relation to their spatial environment means being aware of how global and international flows are assimilated into local construction. In other words, all urban areas in development, such as New Towns, must be involved in the process of defining identity.

The different identities of a place – including those associated with social and cultural environments, urban and regional knowledge competencies and various spatial features of a city or region – need to be taken into account in the development and renewal of spatial environments.

As a result, the basis for development and balance of the territorial identity should be a comprehensive understanding of urban strengths and identities, featuring history and development dynamics. It is also necessary to acknowledge less positive features in a given city or region and to assess how these features can be applied constructively in development.

---

91 Nguyen J. *The challenge of identity for the European New Towns University*, Master degree, Université de Lille I, 2009, p. 81.

92 ACTED, *Culture as a tool for development: challenges of analysis and action*, Paris, 2009.

93 Ruggeri D. "Constructing identity in master planned utopia: the new town of Irvine, CA", Conference communication paper, Alt Empordà, October 2008.

Thus, in the future, one of the major challenges that local authorities will face will be the commitment to a more long-term and sustainable approach to urban development that does not differentiate between centre and periphery, but rather, that is based on a more varied and realistic picture of a place's identity. In other words, this would entail strategies that take into consideration both the negative and constructive aspects of urban identity.

It is also important to note that development strategies must be rooted in the cultural, natural and commercial structure of an area to ensure sustainable development. Urban policies and regional strategies should be structured so as to respect the place and the life led there, in terms of transportation and environmental planning and other aspects relevant to everyday life. Development must counteract social exclusion, meaning that the city and region must be accessible to all and integrate physical, social and cultural accessibility.

Everyone must be entitled to share in the benefits of development, not just in urban areas, but throughout Europe and in the member states and their regions. The central message of the European Spatial Development Perspective was that a framework for polycentric development should be created. We must ensure sustainable and balanced development in big cities and neighboring municipalities like New Towns for the benefit of regional development as a whole. As emphasized in local agenda 21 works and the Aalborg charter, holistic development of cities must draw on local diversity.

The key words are integration, involvement, and participation, if we are to ensure well functioning urban and regional policies. City and regional citizens must co-operate with public and private actors as equal partners. We must promote and support internal co-operation and co-ordination between urban and regional actors and between different cities and regions.

Identity is a fundamental problem for New Towns as they were more or less designed on a clean slate:

> The discourses about identity have exhibited two strands: One strand, (ethnos) stresses the importance of heritage. It focuses on community spirit as formed by intrinsic and coinciding relations of joint ethnic status; language, religion and history. The arguments are emotionally based, binding individuals and communities together in common feelings and inherited cultural values. The other strand, (demos) stresses the importance of the contract. It focuses on general rights and responsibilities of the citizens and society together in joint agreement on civil rights and duties[94].

94 Boje Growth N., "Urban and regional identity challenged by globalization", in *European cities in a global era*, Danish Minister of the Environment, 2002, p. 18.

2.9.2. Lack of Historical and Cultural Heritage

New Towns usually do not benefit from real historical roots, architecture, or landscape structures. It is difficult for residents to identify with spaces that were designed within a very short timespan, and, therefore, pride cannot be easily induced.

Obviously, situations vary considerably across cases. For example, Milton Keynes has incorporated both old villages and river routes in its design, which has helped the New Town to forge identity from tangible elements.

Another problem is architecture and design. Although New Towns embodied modernity, the architecture of the 1970s and 1980s is not appreciated much. They give the impression of soulless residential districts without real unity. In Eastern Europe, this impression is all the more accentuated as prefabricated concrete slabs were used to enable quick and mass construction (plattenbau). The use of a modular construction method largely accounts for their perception as monotonous places with huge districts of dismal architecture[95]. Furthermore, the lack of heritage in New Towns impedes tourism, whereas the presence of tourists can foster inward pride and consolidate the local economy.

2.9.3. The "Dormitory Town" Effect

As illustrated earlier, due to the post-war population boom, housing tends to be the dominating feature of New Towns. Economic activities or soft investments (e.g., in leisure and culture) were not a priority and led to a heavy dependence on central cities; hence, the poor image of "dormitory towns" for commuters where nothing happens day or night.

Although painting a largely accurate picture in the past, this preconceived image is changing as New Towns strive to become more appealing. Recent studies in France[96] have shown that residents on a whole are rather satisfied with their environment and the provision of public services, particularly as in regard to sport and cultural facilities. But despite government incentives to create business centres and employment poles, the image of monotonous and unwelcoming towns persists in the outside world, creating a discrepancy between inward and outward perceptions[97]. Indeed, a local leader from the French Town of Saint-Quentin-en-Yvelines explained in 2005:

---

95 Molster A., "New Towns revisited", in *'SCAPE*, 2006.

96 *Les villes nouvelles adoptées par leurs habitants*, IAURIF, 2005.

97 ENTP, *The top 8 challenges for urban regeneration in New Towns*, 2005.

> Our image is good with the people who live and work here, but in the outside world it is bad, or just not well-known. We need to improve our image in the eyes of the outside world.

Even the strong identity inputs by central governments have failed to change this overall image. In France, for instance, the identity of New Towns is closely linked to education thanks to the presence of universities, but this has not been enough to create a dynamic environment for young people. Moreover, towns such as Cergy-Pontoise (France) cite negative side effects generated by the presence of students; more specifically, the high share of cheap student housing means that it is difficult to attract middle-class households. According to most New Towns, increasing social diversity is a necessary requirement for improving their overall image.

2.9.4. *Lack of Community Spirit*

Community spirit and sense of belonging were, and still are, difficult to achieve in New Towns. This is mostly due to their rapid growth from scratch, but also to the fact that the first residents tended to move to these towns out of rational (not emotional) reasons[98]. In other words, they decided to do so primarily because of the availability of affordable housing – not because there were job opportunities or a renowned cultural heritage. Consequently, they identify with their homes rather than with their towns. However, the rise of a second generation creates space for future prospects but also makes forging an individual identity all the more urgent. Furthermore, a lack of pride can indeed lead to safety and security problems[99]. For instance, as reported by the economic officer of the British New Town of Craivagon in 2005: "The lack of identity and the lack of pride raise problems with antisocial behaviour".

2.9.5. *Filling in the Gap*

Piecing together a common identity is one of the most decisive challenges for New Towns. This requires a long-term development vision shared by all stakeholders and the population – not imposed, but through an inclusionary and participatory process.

2.9.5.1. New Towns and Branding Strategies

To improve their overall image, should New Towns suppress their special nature as modern and planned communities or claim it as a defining feature people should be proud of? It seems unreasonable to

---

98 Molster A., 2006, *op. cit.*

99 ENTP, *The top 8 challenges for urban regeneration in New Towns*, 2005 http://www.newtowns.net/themes.

think that New Towns will become mainstream urban entities anytime soon, as their physical infrastructures bear the imprint of a clearly identifiable period. Besides, rejecting their nature would automatically question the very reason why they were created in the first place and, thus, their whole existence. On the contrary, spotlighting their unique attributes and numerous achievements through communication tools could have a positive impact on their outward image.

The aim would be to turn New Towns' supposed weaknesses into assets, enhancing, for instance, their young population and visionary nature[100]. Sticking to one branding strategy would contribute to inspiring a sense of pride among residents, which could subsequently act as a force in the development of community spirit.

Branding strategies also raise the question of physical infrastructures: what could be done to counter citizens' negative perceptions and to cultivate their sense of belonging to the city? Regeneration has obviously become necessary in most of the New Towns because of ageing buildings and public spaces. But once again, it would be a shame to lose all distinctive features through a process of destruction and rebuilding. Measures could thus be undertaken to preserve the New Towns' design and to modernize infrastructures[101] to improve their functionality and adaptability to forthcoming challenges. Also, helping citizens to see their living environment with fresh eyes is critical to changing the image of New Towns. Therefore, it is necessary to promote the understanding that architectural and urban planning choices embody an important period in history[102].

---

100 *Ibidem.*

101 See the previous best practice on Hässelby-Vällingby.

102 ENTP, *The top 8 specific challenges for social cohesion in New Towns*, 2005 http://www.newtowns.net/themes.

Case study: Our Basildon (United Kingdom)

– "Our Basildon" is an outwardly simple project designed to **promote community cohesion and sense of belonging in the New Town**. It is based on a 12 km trail that weaves in a figure of eight pattern through the centre of Basildon, showing the main elements composing local heritage (e.g., landmark buildings, figurative and abstract sculptures, etc.). Although Basildon claims its identity as a New Town, loud and clear, it also promotes ancient remains through this initiative and thus a more profound urban history.

– The trail was opened in 2009 with the **following aims and outcomes**:

- Increasing local awareness of the built heritage and the open spaces
- Increasing tourism to the area
- Using the initiative as a way to involve people and boost economic regeneration
- Developing the trail concept and promote it as an exemplary project for other towns

– **What are the replication possibilities?** "Our Basildon" is an innovative project that has the potential to showcase Basildon in a positive light. It could be a source of inspiration, but:

- Initiators have to ensure that the local heritage can be enhanced and rediscovered through modern works of art, exhibitions or other events. Basildon stresses in its communication strategy the diversity of its heritage and hence its interest for both residents and visitors.
- The involvement of the population is a crucial factor in fostering a sense of community belonging. For example, Basildon has involved children in the process of piloting trail project. Otherwise, the population will not feel concerned by the project, but excluded instead (e.g., top-down approach).
- The project must be conceived at a larger scale as a way to foster development, as shown by the projects resulting from the partnership between Basildon and the NHS South West Essex (e.g., healthy eating initiative or installation of outside exercise stations along the trail).

2.9.5.2. New Towns' Imaging and Icons

At a time when mass-media enables people (i.e., potential tourists, inhabitants, businesses or investors) to widen their knowledge about the world and what's new out there, the issue of image has fully become a part of urban projects and competitiveness policies in general. In the case of New Towns, the creation of strong images and icons seems to be a powerful lever of action to help resolve the monotonous stigma of districts or the lack of public spaces for people to meet and enjoy the outdoors.

The development of cultural and leisure activities could be one way to achieve this goal. In many European countries it has been observed that culture plays a significant role in the development of a city and a region[103]. Although the extent to which it has direct positive effects on the economy is not fully determined, it is clear that it creates jobs and influences other sectors, such as accommodation or shopping. Culture also has a qualitative effect on the town's image, civic pride and identity, in so far as it enables the possibility to change the preconceived notion that "nothing happens" in New Towns. It is also a way for towns to maintain and enhance their local culture while drawing inspiration from global trends. This is why New Towns tend to invest in leisure and cultural activities; they want to provide services that will keep residents busy in their own hometown (rather than escaping to the more exciting big city). Construction or renovation of shopping malls, leisure centres or museums are symbols of this will, but the effect must address local needs.

---

103 Helforoush R., "Fantasy city and identity in new towns", in *International Conference on New Towns Report*, Ministry of Housing and Urban Development of Iran, Tehran, 2005, p. 462.

"Thanks to Europe, the town will lead economic, social, touristic and environmental projects…" Use of a "European image" by a French New Town (2005)

Town image is materialized through the development of landmark buildings or sites which are supposed to reflect not only current ambitions and trends but also the cities' future vision. Municipalities tend to reconcile social needs with beautification of public spaces, which is why new public facilities are often privileged. They enable municipalities to adapt to the changing (and often growing) demand while recasting the urban tissue as more bold or innovative.

Some New Towns enjoy a strong external image, even though it may not always be positive. For instance the city of Marne la Vallée, divided into several sectors, has its image strongly anchored in "Eurodisney Park", which is located in the last district. The presence of a strong university, "Cité Descartes" and, more recently, that of a sustainable renovation of buildings cluster may alter the image to reflect a more dynamic and knowledge-based region.

Almere and Lelystad (the Netherlands) are strongly linked with the image of polders, as they occupy land reclaimed from the sea. The huge construction undertakings in Almere – e.g., housing areas with pavilions (Muzikwijk, etc.) and the town centre – have bestowed the image of a city where anything can happen. The construction of a bridge connecting the town to Amsterdam has reinforced Almere's futuristic image. The town's slogan is "Het kan in Almere" or "Anything is possible in Almere".

The slogan, "Almere for the Almere people" or "Almere voor de Almerders' may be ironic for those who know the history of this town, where most inhabitants are recent migrants. However, Almere's image has been dampened by the results of the 2010 local elections, which saw the extreme right[104] gain power. This transformation of image – from "future towns", to "dormitory towns" to "windows of the extreme right" – demonstrates that identities are not static. Turning them into assets will be a real challenge for New Towns because identities certainly need to be "nurtured, safeguarded, and remolded" to impact on outside image[105].

2.9.5.3. From New Towns to Eco-towns?

Many New Towns have understood the importance of new environmental constraints and are striving to revive their initial image as pioneers in quality of life. Cergy-Pontoise (France) has initiated a sustainable development approach to enhance the territory and reinforce solidarity among its city council members and population. For instance, 370 square meters of photovoltaic panels were installed on the roof of the main administrative building. The aim is to build on this experience

---

[104] Partij Voor de Vrijheid: PVV party from Wilders which scores first at the local elections of 2010.

[105] Scheffler N., Kulikauskas P., Barreiro F., "Managing urban identities: aim or tool of urban regeneration?", in *URBACT Tribune*, 2009, p. 13.

within the Agenda 21 framework and to install photovoltaic panels for every re-roofing when possible. Another goal is to implement energy and water consumption monitoring systems in certain buildings. The future geothermal boiler in Val Maubuée will significantly reinforce the town's commitment to sustainable development; its bold design also represents the opportunity to create a flagship building.

Geothermic boiler in Marne la Vallée-Val Maubuée

2.9.6.4. New Towns Need Citizen Participation

As the notions of "governance" and "participatory democracy" become prevalent in public debates, participation is an increasingly important aspect of political life. The particular way in which New Towns were developed underscores this. Indeed, they were planned according to a strict approach devoid of resident input; buildings preceded inhabitants and the emphasis was set on developing the town's physical structure. Changing social demands now require a different approach.

The bonds between citizens and public authorities should be strengthened in order to reinforce understanding and endow inhabitants with a sense of responsibility[106].

Whether New Towns have succeeded in becoming "real" towns after 30 or 40 years is related to their capacity to gather inhabitants around common interests. Studies[107] have shown that civil society groups have played a key role in town planning and management. In the case of Val

[106] ENTP, *The top 8 specific challenges for social cohesion in New Towns*, 2005 http://www.newtowns.net/themes.

[107] Tayab Y., "Le rôle des différents acteurs dans l'aménagement et la gestion des villes nouvelles en France: l'exemple de Marne-la-Vallée", in *International Conference on New Towns Report*, Ministry of Housing and Urban Development of Iran, Tehran, 2005, p. 395.

Maubuée (France), the existence of 1,400 associations in 1994 had enabled inhabitants to express their views and interests, compensating for the lack of convivial places. Even though civil society groups do not have the power to block or affect planning projects directly, local representatives are aware of their influence and take their concerns into account (e.g. proposals made within the framework of the town's master plan)[108].

Achieving a shared sense of belonging, however, will require many pro-active initiatives. The promotion of community involvement will have to be underpinned by major participatory initiatives – ones enabling citizens to do more than just hear information or make suggestions. Similarly, volunteer work should be enhanced and encouraged because it contributes positively to the developing social infrastructure of New Towns. Some New Towns have already implemented successful and innovative participation initiatives; they increasingly use the information and communication technologies as a flexible and appealing tool for citizens as demonstrated through Harlow's Hotspots initiative.

Case study: Harlow Hotspots (United Kingdom)

– With the help of Harlow Star, Essex County Council and Harlow Council, Harlow Renaissance has established **an annual competition giving free rein to citizens' views**. Community groups, societies and tenant and voluntary associations throughout the town were invited to nominate a project that would enhance their local environment for the benefit of all the community. Then citizens were free to vote for their favourite project.
– For the 2008 competition, 19 nominations were shortlisted and the winners were the Parndon wood and Hawkenbury meadow, as well as the new sculpture for the town park paddling pool. Competition organisers, Harlow Renaissance and Essex County Council have put up £5,000 each to pay for the work (e.g. removal of trees, replacement of a pathway, etc.).
– The **aims and outcomes** of the project were to:

- Foster participatory democracy and sense of community
- See the New Town through fresh eyes

– **What are the replication possibilities?** "Harlow Hotspots" is a simple budget participatory initiative that can lead to fruitful results for citizens. Indeed, one can expect that the focus on local issues will increase the inhabitants' sense of responsibility for their hometown. Although the scale of the projects are rather limited, the positive impact (from low-costs and citizen involvement) is potentially strong. Thus, the replication of such a good practice should be easy, but municipalities should nonetheless see to it that a wide range of groups are represented.

[108] *Ibidem.*

2.9.5.5. *New Towns and Culture: a New "Renaissance"?*

Another way for New Towns to improve their image, while enhancing sense of community and social cohesion, is to promote local cultural activities and facilities. In addition to the fact that culture can be used to attract high-income groups, it can also cultivate interactions between people from very different backgrounds, thus fostering social cohesion[109].

Moreover, the renewed status of culture is the corollary of another structural change in the nature of the European city. Indeed, 50 years ago, cities were critical centres for trade, industries and other economic activités, whereas nowadays they are increasingly dedicated to providing entertainment and cultural activities. This may help to explain why – as illustrated earlier in the case of the Hergé Museum – New Towns try to provide more cultural facilities and a wider spectrum of activities.

Several successful examples demonstrate that New Towns can use their potential as "pilot cities" to promote a good outer image and to foster a sense of community. For example, under the slogan "Yes we can", Almere (the Netherlands) launched the "Almere 2018" initiative in order to be selected as the European Capital of Culture in 2018. In the project summary, "Almere 2009 – 2018: Shifting toward a cultural city", the town clearly links this event with its desire to complete its "integration phase" and become a mature and sustainable town, with a strong cultural identity, able to tackle the challenges raised by migration and the lack of social cohesion[110].

---

[109] European New Towns Platform, *The Top 8 Specific Challenges for Social Cohesion in New Towns*, Brussels, 2005, p. 4.

[110] "Almere 2009 – 2018: Shifting towards a cultural city" on the website of the bid: http://www.almere2018.eu/home_en.

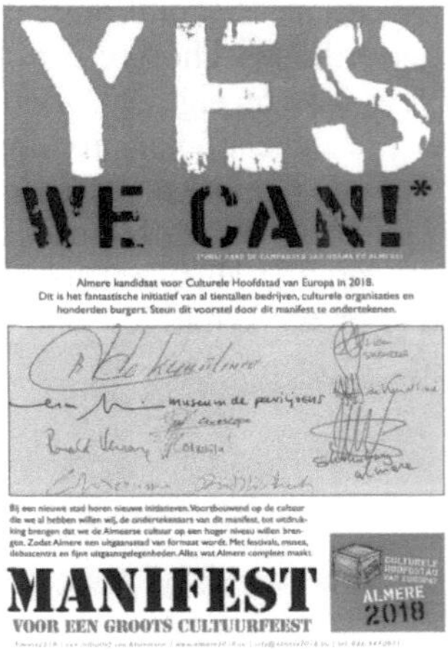

Manifesto for Almere cultural city in 2018

Almere's campaign to become a cultural city[111]

The example of Cergy Pontoise (France) shows that New Towns can also attract cultural infrastructure by taking advantage of their existing assets and particular characteristics. Indeed, Cergy Pontoise recently won the call for proposals launched by the French Ministry of Culture and Communication in order to select the site of the future National

[111] http://www.almere2018.eu/info.

Centre of Heritage Management, Restoration and Research of the Ile de France region. The French Ministry chose Cergy Pontoise because of its special features as a New Town: e.g., its good connections with the mother city (Paris), the presence of various universities and research centres, its good quality of life (lots of green spaces, public services and shopping and cultural facilities), identifiable, institutional stakeholders and unimpeded use of the land[112]. This centre has the potential to give a boost to cultural life and to create synergies with the local stakeholders.

[112] Communauté d'agglomération Cergy-Pontoise Energies Ouest, *Synthèse – Centre national de conservation, de restauration et de* recherche *du patrimoine*, 2009.

3. Town Centres

Town centres are the hearts of cities. Ideally, town centres should be hosts of urban economic growth; such strategic places of business and commercial activity need to be more visible. Town centres are also the pride of politicians that want to promote town image. Finally, town centres are places for investors and represent public spaces and meeting points for inhabitants. Town centres should be creative places for living, economic activity and planning. Built with the idea to become polycentric cities, New Towns often lack a real town centre. Some New Towns are in the process of creating a new centre (e.g. "Le carré Sénart", a centre designed in the form of a square).

3.1. State of the Art

Town centres experience difficulties linked to economic revitalization, accessibility and ageing/maintenance of buildings. This is jeopardizing their role as the heart and image of cities. They need to be vibrant communities with commercial areas and a solid social nightlife economy. The objectives of the town centres are also to reflect the image and identity of a town and to be part of the spatial development of local and regional areas. Finally, they need to function as central nodes of activity, connecting different urban networks and acting as the crossroads for economic exchange, communities and inhabitants[1].

3.1.1. Traditional Role and Transformations

Town centres are "stereotyped" as ideal and central places – "agoras", "crossing and meeting points", "places for culture", as exemplified by the presence of religious monuments[2].

In reality, however, they now face many problems with deprived areas. Town centres experience all kinds of transformations because they must adapt to changes in lifestyle and the economy. Nevertheless, New Towns experienced many transformations in the past which can be drawn upon for addressing weaknesses and opportunities. Before the development of car access, the stereotypical image was that European

---

1 We could also add the knowledge production and transfer with the presence of schools, universities, libraries and museums.

2 This used to be the case not only in Europe but also in other continents' traditional towns.

cities and towns had a unique town centre concentrating all functions and activities. Town centres were indeed central places for retail, inhabitants, economic and administrative life, accessible by foot since the size of cities and towns were relatively limited and compact. The rural exodus of the 19th century resulted in the congestion of large city centres, which suffered from overcrowding and high concentrations of the urban poor (contributing to health and security problems). Towns expanded in an anarchic way, involving an exodus of the middle and upper classes (bourgeoisie) to the suburbs – the result of which was a detrimental effect on the town centre economy[3].

With the development of car access came the increasing decline of town centres[4]. Historical centres were not fit for car traffic and parking space were lacking. Thus, traditional town centres suffered from congestion and numerous economic activities moved closer to communication and transportation hubs (train stations etc.) "As the traditional centre loses the benefits of functional agglomeration and its primacy in the transport network, what happened is the dissolution of the central city as a political, economic, social and symbolic locus[5]".

Towns like New Towns became polycentric with several monofunctional centres of activity. In order to tackle new social, economic and environmental challenges, a renewed interest in city centres as hubs of both commercial and residential activity has emerged in the last decades[6].

As a result, policy documents in Europe (e.g., UK, France, Netherlands, etc.) have increasingly sought to emphasize the role of town centres as appropriate locations for residential, retail, cultural and leisure development; therefore, supporting a functional/social mix and

---

3 Claval P. "De la logique des villes aux ressorts de l'urbanité", in Da Cunha, A. & Mathey, L. *La ville et l'urbain: des savoirs émergents*, Months, Presses Polytechniques et Universitaires Romandes, 2007, p. 64.

4 City development also became less and less connected to natural/geographical/ topographical aspects with the idea of progress.

5 Oc T. and Tiesdell "The death and life of city centres", in T. Oc. Tiesdell Safer city centres: *Reviving the Public Realm*, London: Paul Chapman Publishing Ltd, 1997, p. 15.

6 Ravenscroft N., "The vitality and viability of town centres", in *Urban studies*, No. 37, 2000, p. 2533-2549; Donzelot F., "La ville à trois vitesses: relégation, péri urbanisation, gentrification", in *Revue Esprit*, Paris Mars-avril, 2004, p. 14-40; Oc T. and Tiesdell "The death and life of city centres", in T. Oc. Tiesdell *Safer city centres: Reviving the Public Realm*, London: Paul Chapman Publishing Ltd, 1997, p. 1-21; Smith N. and Williams P. *Gentrification of the city*, Allen & Unwin, Boston 1986, p. 35-55, O'Connor J. and Wynne D. "From the margins to the centre: cultural production and consumption in the post-industrial city", *Working Papers in Popular Cultural Studies* No. 7, 1993.

rehabilitation approach to city centres (vs. Tabula rasa concept) and fighting against urban sprawl (largely inspired by the 1994 Aalborg Charter and its promotion of Urban Sustainable Development)[7]. Nowadays, while historical cities show a renewed interest in their centres as places of main attraction, other peri-urban towns with less cultural heritage face even more difficulties in making town centres enjoyable places for inhabitants. This is an asset for New Towns.

3.1.2. Today's Situation and Difficulties

In general town centres are still transforming but we can learn from the past in terms of urban planning and design. Town centres have also undergone a "recentralization process" by gaining interest from inhabitants[8].

Throughout this process, many inhabitants who had previously left city centres (for the suburbs) are now returning; in some large cities, this had led to a phenomenon of gentrification of inner city residential neighbourhoods[9].

This renewed interest in town centres is also the result of the entrepreneurial exploitation of culture, consumption and town centre image; in reality, however, there has only been partial revitalization and regeneration of these centres[10].

More and more, local authorities need to adapt to societal and economic situations. Anticipating changes linked to demography, life style, the economy and the environment remains a challenge. Town centres are once again embraced as the heart of cities. But in a context of economic crisis they face difficulties in developing, regenerating and maintaining buildings, facilities, and public spaces. Since the end of the 1990s, there is a new emphasis on town centres appearing in the policies and regulations of European countries[11]. This includes the understanding that town centres need to be managed in such a way as to incorporate the strategic development of both private and public areas. This revived interest in the management of town centres, as expressed in recent policy documents, emphasizes the need for partnerships and collective

7 Cyria Emilianoff in http://www.developpement.durable.sciences-po.fr/publications/cahier8.pdf.

8 O'Connor J. and Wynne D "From the margins to the centre: cultural production and consumption in the post-industrial city", *Working Papers in Popular Cultural Studies* No. 7, 1993.

9 Donzelot F. "La ville à trois vitesses: relégation, péri urbanisation, gentrification", in *Revue Esprit*, Paris Mars-avril, 2004.

10 Smith N. and Williams P. *Gentrification of the city*, Allen & Unwin, Boston 1986, p. 35-55.

11 Ravenscroft, 2000, *op. cit.*, p. 2533.

action[12]. Thus, the main purpose of town centre management is to influence the organization of space, covering public and private areas and interests.

Lifestyle and economic changes in town centres also requires a new set of needs, in terms of infrastructure, public services, housing, transportation networks and commercial facilities. Dwindling public financial resources call for more public-private partnerships and further economic revitalization of town centres.

3.2 Specific Challenges Facing New Town Centres

New Town centres tend to face the same problems (e.g., social, economic, need of regeneration, accessibility, etc.) that emerged in the absence of a sense of community. They often lack a proper town centre or/and connectivity between different centres.

The challenges faced by these town centres are grouped under three themes: mono- functionality/segregation (theme 1), economic revitalization (theme 2) and accessibility (theme 3).

3.2.1. Theme 1: Zoning and Segregation

Zoning is an important issue for New Towns; business areas and housing settlements are clearly separated from each other. A strict separation also prevails in housing areas, e.g., between social housing and private housing units. This results in segregation of city centres and in the emergence of new "ghettos". Thus, city centres suffer from decline, as evidenced by the need to regenerate ageing buildings. Retail and services are more important than in other areas of town, but most inhabitants venture to the town centre usually by car and mainly to do evening and weekend shopping. Some town centres are host to a high concentration of social housing, as much as 80% in some town centres. In other towns, the centre is mostly reserved for retail. There is now a priority to change this mono-functional pattern and to create mixed use areas in order to (among other things) create a better sense of safety and security in the evenings. In the regeneration of town centres; the goal is to mix areas at the social (mainly by drawing the middle class to the town centre) and economic level, as well as to develop a mix of activities (e.g., cultural, leisure, etc.) and public spaces.

The idea of "centrality" in New Towns was paradoxical from the start – jeopardized by the co-existence of several burgs and villages, each with their own mono-functional centre, ultimately resulting in the phenomenon of polycentrism.

---

12 *Ibidem.*

Polycentricism had different impacts on segregation; it led to the emergence of ethnic areas or ghettos, but also encouraged a more social mix in collective areas. The urban transformations have left deep marks that concern all aspects of society, from mobility to the aesthetic quality of buildings to our ways of life (as citizens, families, consumers, employers or workers). Accessibility problems had direct impacts on social patterns (zoning and segregation).

In most European suburbs and in all project partner towns, housing was designed for nuclear families.

This situation is now changing in the industrialised world. Local authorities face two important demographic processes: (1) foreign migrants with large families and low income groups in need of decent housing (2) and population ageing. Such changes not only require policy strategies focused on fulfilling future housing needs, but also those able to deliver suitable transportation, services, leisure activities and other facilities.

Finally, it is possible to list the different zoning problems experienced by the New Towns:

- Zoning of the town centre
- Difficulty with access to public utilities/ lack of public spaces and facilities
- Lack of parking spaces
- Deprived areas and public spaces
- A population with social difficulties
- Low attractiveness of the centre of retail

Above: Shopping mall in Emmen, the Netherlands
Left: Town centre in Emmen, the Netherlands
Below: shopping mall in Milton-Keynes[13]

3.2.2. *Theme 2: Image and Economic Development*

The image and attractiveness of New Towns have declined because of derelict areas and the absence of town centres. They are no longer considered the 'best choice' for those who want to live outside the congested urban areas. People initially moved outside of large agglomerations for higher environmental quality and lower land prices. Policies designed to attract the middle classes through the provision of better housing and education services have now changed. Without real town centres or with town centres in need of regeneration, it is difficult for the project partners to find a real identity. Rather than becoming real communities, they were simply places to live and are now generally regarded as dormitory towns, despite the fact that employment has grown.

[13] Picture taken by the author, 2009.

Jobs have been created in the towns' centres mainly because of their close proximity to large cities. They have attracted businesses by offering good infrastructure, tax incentives and facilities. The town centres also need further development of cultural and leisure activities. The obstacles to create new mixed areas are the low densities of town centres, the land use properties that are divided among numerous owners and the difficulties with economic activities in the centres. The economic crises also directly impact the life of town centres (i.e., declining shopping areas, the credit crunch effect on small retails, etc.). New policies intend to steer public and commercial activities to the centre. To achieve this, they need to promote the town centre and to reinforce trust and confidence among businesses, investors, city developers and inhabitants. For some of the partners, the development of large retail units will attract regional economic activity. For other partners, the development of leisure opportunities and cultural facilities in order to improve the image and identity of town centres is also a challenge. This is why the municipalities need to develop tools:

- To consolidate and reinforce the retail with the expansion of floor space;
- To deliver a high quality public atmosphere by ensuring safe, attractive, and lively streets and places;
- To expand the mix of activities in the town centre, including leisure opportunities, residential development, cultural activities and, where appropriate, smaller scale office space and commercial activities;

To co-ordinate strategies for the marketing, management and maintenance of the town centre around consistent messaging that enhances image.

3.2.3. Theme 3: Accessibility of Town Centres

The town centres of all New towns experience accessibility problems. At the time of rapid urbanization some decades ago, transport provision was a critical issue because of the long distances to the centre of the major city, but low densities of these new urban areas made it difficult to build functional and efficient infrastructure.

Both in New Towns and in the periphery, infrastructure planning has not caught up with urban development. In the best cases, the demand for transport between the new urban areas and the original town centres has been met. While most partners' master plans included provisions for lateral connections with other centres, the available budgets are generally insufficient to deliver the planned infrastructure. The public transport links that have been built link the centre of the metropolitan area to the periphery via rapid transit, metro and motorways. Internal rail, bus routes and links with other areas in the periphery were not provided; therefore, virtually all of these journeys are made by car. Most of the town centres were built in a period when car use was increasingly popular and on the rise. This impacted today's urban structure; partner towns have a lack of public transport connections to and from the town centres. In some towns, the streets are too narrow to permit bus accessibility. This is comparable to New Town centres that are usually pedestrian areas only accessible by car.

Nowadays, most New Towns have a public transport policy which includes the town centres; many even propose better bus or light rail connections, high-quality facilities and frequent services. In some towns, improving public transport will require considerable investment in order to overcome the problems posed by the current urban design. In other cases, as in France, local authorities are relying on regional or national decisions to improve the transport system. The transport policies of most towns are still quite limited; car use is still facilitated by providing parking spaces, and many areas in the partner towns are not accessible by public transport.

Increasing urban densities may be an important condition for the introduction and use of more sustainable transportation systems. This is a longer-term goal and will not be easy to achieve, for it would mean developing real town centres and require further economic growth along with higher local authority budgets. In the meantime, decision makers have realised two things: (1) increasing congestion is not an asset for their town and (2) they need an alternative public transport system directed primarily at the elderly, lower-income families and students. They also need to offer viable options for people who do not want to travel by car everyday. Mobility and transport have implications for all other aspects of the partner towns, such as their economies, urban planning and governance.

Traffic congestion, lack of identity and increasing land and rental prices are just some of the problems affecting the regeneration of town centres. Cuts to national government budgets have exacerbated the situation and are obstacles to the regeneration of town centres.

In town centres, some buildings are rapidly deteriorating and not in accordance with today's environmental, aesthetical and health standards. They have difficulty attracting people, businesses and the middle classes, reducing possibilities for further development.

This is why New Towns need to identify tools to:

– Better connect the town centres to surrounding areas;

- Ensure inter-modal transport in order to increase attractiveness of the centre;
- Develop high quality access to the town centre and within the town centre by a variety of transport modes, maximising opportunities for sustainable forms of transport.

Town centres have to be competitive and to attract retail because they create jobs, enhance economic activities and are best positioned to respond to the Europe 2020 objectives. New Towns face the challenge of consolidating and restructuring the urban fabric to accommodate changes in the population and in lifestyles, as well as to redefine their sustainability objectives.

At the European level, numerous tools and policies have been developed to promote an integrated approach to urban development.

The 1998 "Action framework for sustainable urban development in the European Union" established the following objectives for the rehabilitation of urban zones:

- To aid the long term unemployed, young people in a situation of exclusion, and other appropriate groups;
- To stimulate economic development and employment;
- To improve the urban environment;
- To prevent urban crime;
- To strengthen local capacities in order to find integrated solutions.

The European Union developed numerous initiatives like the URBAN program, which in the second period represented 700 million Euros spread across 70 urban districts experiencing difficulty.

In 2007, ministers also signed in Leipzig a charter giving recommendations on the development of sustainable European cities, namely the following:

- Making greater use of integrated urban development policy approaches;
- Creating high quality public spaces;
- Modernizing infrastructure networks;
- Giving attention to deprived neighbourhoods in the context of the city as a whole;
- Pursuing strategies for upgrading the physical environment;
- Strengthening the local economy and local labour market policy;
- Promoting efficient and affordable urban transport.

In terms of image, indicators have been developed by European universities to evaluate the vitality and viability of town centres[14]:

While town centres in large cities flourished thanks to their renewed attractiveness and gentrification, the town centres of peri urban areas (New Towns and regional growth poles) were neglected.

The creation of large retail stores and places of leisure at the periphery of such cities, in particular, has decreased the attractiveness of town centres. This underscores the need to develop integrated policies which better reflect today's world.

---

[14] Ravenscroft N., 2000, *op. cit.*, p. 2537.

Conclusion

The aim of this book was to gather theoretical and empirical elements in order to draw an extended and in depth comparative analysis of European New Towns, with a specific focus on France, the United Kingdom and the Netherlands, but also taking into account experiences from Northern, Eastern and Central European countries and from "emerging" countries in other world regions. Through cross-cutting themes and examples, the analysis concentrated on New Towns undergoing a transition to "Mature" Towns and endeavored to answer a fundamental question: what are the main challenges for New Towns today and how can they cope with them?

After briefly describing the policies that enabled their creation in the 20th century, the emphasis was turned to the main issues and challenges faced by New Towns today:

Scarce financial resources: Most New Towns feel the increasing pressure of self-financing due to the state's disengagement; it is sometimes difficult to attract investment, although a great deal of work has been carried out to enable a better balance between jobs and housing.

Shortfalls in housing and services: Even though a great deal of progress has been made in recent years, public amenities remain a sensitive issue and must be properly addressed in order to provide a better quality of life for citizens and to help them establish more lasting roots in their living environment. Both the housing stock and infrastructure are in acute need of urban renewal due to their specific history.

Lack of identity and sense of community: This is a fundamental and cross-cutting issue that requires specific attention. Indeed, it wields a deeply felt influence over the external image of New Towns, which in turn impacts the decisions of investors and potential inhabitants. At a time when growth is stagnating or even decreasing, rapid action is called for to improve their overall image. This important point reminds us, however, that image and identity is not just about technical engineering, but that it has mostly to do with representations and imaginaries. Does this mean that municipalities should employ "dreamers' in their municipal planning departments? Maybe not, but artists and cultural vitality create imaginaries that will cultivate image.

As it has been extensively shown, New Towns are criticized today because they tend to be mono-functional and are not "environmental friendly" enough. Yet, let us not forget that New Towns were creative laboratories and the lessons drawn from their experience are still relevant for the new development policies of "satellite towns" in other countries.

Moreover, the case studies recounted in this book underline the fact that, in terms of economic development or social cohesion, New Towns have no reason to be envious of "old" medium-sized towns. Therefore, it seems that the renewal of the New Town mindset – advocated by many urban planners, New Towns and reports – is being increasingly emphasized. The evaluation grid following this conclusion was designed as a tool to evaluate New Town policies and to determine their strengths and weaknesses.

Set of Indicators: Successful and Sustainable New Towns

DIMENSIONS	OBJECTIVES (WHAT?)	IDENTIFIED PROBLEMS (WHY?)	POSSIBLE SOLUTIONS (HOW?)
MINDSET: Better manage the development of New Towns	Analyze the current local situation: – Define the existing and ongoing actions – Identify the needs	– Indicators are missing/not good – Lack of integration of certain stakeholders	– Detailed SWOT analysis – Cooperation with stakeholders; participatory democracy
	Generate a clear vision for the future: – Define a long-term ambition – Define the means to reach the ambition	– Driving ambition not clearly defined/not realistic – Insufficiently structured and shared approach – Ambition not matching the local context – Lack of capacity to deal with changes	– Detailed SWOT analysis – Strategic approach based on the cooperation with stakeholders – Use of local assets and potential – Flexible, adaptable planning system
ECONOMIC: Pursue and consolidate the growth of New Towns	Reinforce local competitiveness: – Improve the quality of the labor force – Focus on fields of excellence – Enlarge the scope	– Lack of skilled labor force – Lack of distinctive asset activities – Lack of diversity; hence sensitivity to conjuncture	– Invest in education system and/or attract high added-value activities – Prioritizing investment in identified fields of excellence (e.g. research and innovative activities) – Diversification strategies based on local resources
	Attract activities to the territory: – Reinforce accessibility – Provide good implementing	– Insufficient/inefficient transport system – Bad integration/lack of coherence with the wider region – Local tax system lacking incentives	– Clear ambitions in terms of geographical scale – Investment in infrastructure and multi-scale services – Cooperation with relevant lo-

	conditions – Develop commercial space – Foster (business/leisure) tourism	– Lack of centrality – Dependence on shopping centres – Little attraction for tourists	cal/regional/national authorities – Tax incentives, attractive local fiscal policy – Propose new concentrations of activity – Develop retail stores – Planning projects – Communication campaigns
SOCIAL: Foster social cohesion among the inhabitants of New Towns	Ensure provision and diversity: – In housing – In facilities – In spaces	– Economic growth not followed by an increase in the housing stock – Similarities and monotony of housing estates – Lack of population diversity on the economic level and overrepresentation of low-income households – Under-provision in some fields (e.g. leisure) – Under-provision for specific targets (e.g. elderly) – Lack of diversity and proper "zoning" of urban functions	– Housing projects that respond to demographic growth – Resort to different designs and types of architecture – Providing the widest range possible of housing types and prices – Attracting middle-class through leisure, culture… – Clear understanding of flaws and adaptation measures through long term loans, regional/national subsidies… – Multi-functional integrated spaces
	Foster a sense of community belonging: – Increase participation and encourage social involvement – Improve the quality of public spaces – Develop availability and accessibility to cultural activities	– Little participation capacity in the decision-making process – Culture of participatory democracy not widespread – Insufficient/unused/deteriorating public spaces – Inhabitants unable to identify with town and public spaces – Uncertain return on the investment – Access to a privileged part of the population	– Develop citizens' initiatives to reinforce community harmony and belonging – Enhance the culture: initiate if needed differentiated policies according to the target groups… – High quality/iconic public spaces – Use of lower quality or under-used land – Target middle-class – Tariff policy adapted to the local context

ENVIRONMENTAL: **Reinforce** **the attraction of the** **New Town**	<u>Guarantee a high quality of life</u>: – Improve the durability of the architecture – Reinforce the efficiency and service of the transportation system – Increase the presence of nature	– Bad building performance (e.g. energy), hence the necessity to renovate/destroy fast – High dependency on cars – Absence of less pollutant public transport – Low rate of walking/cycling – Low availability of open spaces – Nuisances: lack of biodiversity, pollution – Bad integration of nature in the city	– Communication campaigns – Increase both planning control and incentives to change cultural practices – Environmental-friendly transport strategy and/or international commitments – Development of alternative ways (e.g. tramway) – Development of pedestrian spaces and cycling paths – Green and blue plans (e.g. green corridors, rehabilitation of river banks) – Differentiated management of green spaces – Mixing urban parks and buildings
	<u>Seize the economic opportunity</u>: – Feature pilot projects (e.g. eco-quarters) – Shift to eco-business	– Showcase effect instead of benefits for the population – Social similarity of the population (upper-class) – Economy threatening the environment, i.e., competitiveness vs. sustainability – Lack of resources – Fashionable trend that might be temporary	– Think social welfare first – Provide widest range possible of housing prices to guarantee social diversity and accessibility to the new area to low-income households – Transversal/integrated approach to environmental issues – Investment at any scale: heating system, eco-technologies cluster, local production of renewable energies… – Avoid mono-specialization

ENTP – European New Towns Platform

The ENTP (European New Towns Platform) is the organization representing New Towns and fast growing suburban centres in Europe. The seat of the organization is in Brussels. Created in 1999, ENTP has become in few years an active network gathering 33 cities and 11 Research institutes from 12 European countries. http://www.newtowns.net. Moreover, the ENTP is currently expanding its extra-European cooperation network. Indeed, we also count among our members a Chinese town (Luodian) and we cooperate with New Towns from Algeria, Morocco, Tunisia and Egypt in the framework of different projects. The ENTP is also working with largest international networks umbrellas in a permanent exchange of good practices to foster more sustainable planning practices and cities.

The main objectives of the organisation are to work as a platform to exchange information, share good practices and to implement common projects. The priorities of the ENTP are: urban regeneration, planning of town centres, housing issues, social cohesion, environment preservation and sustainable development. The ENTP is also a supporting structure of the Covenant of Mayors to promote initiatives to fight against global warming among its members.

Recently, the ENTP participated in the discussion about a political declaration to reduce urban health inequities (WHO Kobe centre), and had the opportunity to review the WHO-UN Habitat "Hidden cities" report.

ENTP activities include the promotion of exchange of information and experiences through seminars, trainings, panels of experts, and the creation of cross-cutting networks of urban planners, sociologists, economists and environmentalists. The ENTP also organises conferences, capacity building actions, research activities, production of material (brochures, leaflets) and workshops. In 2010, ENTP is taking part in a workshop on "low carbon cities" in the framework of the China-Europa forum, and is organising regular working groups and panels of experts in Europe about relevant issues for New Towns (urban planning, environment).

Past and current Experience

Past European projects:

- **Encourage:** Sustainable management of business parks and environmental charters.
- **NEWTASC:** Image and Identity of New Towns.
- **EGENIUS:** Sustainable management of town centres.
- **Migrant Women:** Exchange between European and Latin American local authorities on gender issues.

Publications:

- *Culture in our towns*, ENTP Publications 2006.
- Gaborit P. & Lamas Sanchez P., *New Towns' Challenges in Europe*, European New Towns Platform, Brussels, 2006.
- *News Towns and the crisis, Examples of Good Practices to Tackle the Crisis*, ENTP Publications, Brussels, 2009.
- *Guide to New Towns' Best Practices*, ENTP Publications.

Current projects

- **NEW MEDINA:** From both sides of the Mediterranean from pilot to sustainable towns. This project integrates trainings, expert panels and study visits to promote an integrated approach of sustainable new towns in North African countries.
- **GREENOV:** Sustainable renovation and promotion of energy-efficiency in existing buildings through the creation of green clusters.
- **EUROSCAPE**: Promotion of sustainable management methods of cultural and natural urban and peri-urban landscapes.

Conferences and seminars

- Sustainable cities, Craigavon, Northern Ireland, 2004.
- New Towns image and identity: Tatabanya Hungary 2005.
- Transport in New Towns, Capelle aan den Ijssel, Netherlands 2006.
- The building of a town centre: Sénart, France 2007.
- Harlow: the economic development of a New Town, UK, 2007.
- From New Towns to Maturity: Vantaa and Hässelby Vällingby Sweden and Finland, 2008.
- New towns facing the financial crisis: Alt Emporda, Spain, 2009.
- New Towns in their Metropolises, Agios Stefanos Greece, 2010.
- From Both Sides of the Mediterranean: from Pilot to sustainable towns, Val Maubuée, France, April 2010.

Memberships

- The European Parliament Urban intergroup

The Author and Contributions

Pascaline GABORIT

PHD in political science.

Director of the European New Towns Platform.

Pascaline Gaborit has been the Director of the European New Towns Platform since 2001. In this capacity, She has developed the network. She has organized trainings, events and worked in several European and international projects and research. Her areas of expertise are urban development, social cohesion, gender issues, sustainable development, culture and lifestyles. She participated as expert on different European and international programs and reports linked to urban issues. She completed a PhD in Political Science about "Trust and conflict".

Main publications:

- New Towns image and identities (ed) (2006), Brussels, ENTP publications.
- Genre, temps sociaux et parentés (ed) (2008) Paris, L'Harmattan.
- Les Hommes entre travail et famille (ed) (2007) Paris, L'Harmattan.
- "Reinventing the New Towns Mindset", Nova Terra (2005), September 2005, Delft.
- "De la ville pilote à la ville durable", revue Pouvoirs locaux, Janvier 2004, Paris.
- "New Towns and peri-urbanization", in International New Towns Conference Proceedings, 2005.
- "Culture and development in New Towns: the example of Cergy Pointoise", in ARCADE "Culture and développement"; (2009) Dunkerque.
- Regeneration of urban town centres within New Towns baseline study for the URBACT program, 2008-2009.
- Les stéréotypes de genre: identités, rôles sociaux et politiques publiques (ed) (2009) Paris, L'Harmattan.
- Restaurer la confiance après un conflit civil (2009) Paris, L'Harmattan.

With contributions from

Jonathan NGUYEN has a Bachelor of Arts in English. He completed a double Master's Degree in European Affairs (Lille Institute of Political Studies – Sciences-Po Lille) and European Territorial Development (Lille 1 University). He has written a Master thesis in English about "Imaging New Towns" and graduated in September 2009. He worked on a New Town study for the town of Zoetermeer the same year.

Zoé BUYLE-BODIN completed a Master Degree in European Affairs at the Lille Institute of Political Studies (Sciences-Po Lille). Her background is in Political Science and she is particularly interested in urban planning and sustainability issues.

Reviewers

Eva ZEMANDL is completing a Masters in European Public Policy with the University of Kent's Brussels School of International Studies. Her background is in economic development, social policy, and promoting international business and trade. Her mother tongues are English and Hungarian.

Edoardo GUGLIELMETTI is project coordinator at the ENTP. He graduated in Philosophy in Milan (with a specialization in Mass communication). After a Master in European affairs in Brussels, he worked as assistant for two Members of the European Parliament and worked for the International Training Centre of the International labour Organi–zation.

References

Books

Bouinot J., *La ville competitive: les clés de la nouvelle gestion urbaine*, Paris, Economica, coll Gestion, 2002.

Caldo C., *Geografica umana*, Palumbo, Firenze, 1996.

Da Cunha A., Matthey L. *La ville et l'urbain des savoirs émergents*, Presses polytechniques et universitaires romandes, 2007.

D'Antonio S & Faiella I., *Buone prassi, buone notizie – Le migliori esperienze delle città europee*, Rubbettino editore, 2007, 91pp.

Gaborit P. & Lamas Sanchez P., *New Towns' Challenges in Europe*, European New Towns Platform, Brussels, 2006.

Giddens A., *The Constitution of Society*, Berkeley, Collection University of California Press, 1984.

Hatem F., *Le marketing territorial: principes, méthode et pratiques*, EMS, Les Essentiels de la Gestion, 2007, 292pp.

Ingallina P., *Le projet urbain*, PUF, collection Que Sais-Je?, 2008, 128pp.

ISOCARP – International Society of City and Regional Planners, Low Carbon cities, University of Porto Press, 2009, 261pp.

Jessop B., *The future of the Capitalist State*, Blackwell Publishing, Cambridge, 2002, 344pp.

Merlin P., *Les Villes Nouvelles en France*, PUF, collection Que Sais-Je?, 1991, 127pp.

Merlin P. & Sudarkis M., *From Garden City to Urban Reconstruction: New Towns in Perspective*, INTA Press, 1991.

Mommas H., *City branding: Image building and building images*, NAi Publishers, Rotterdam, 2004.

Reijndorp A., *In Search of a Public Domain*, Amsterdam University press 2006.

Rousset M., *L'action internationale des collectivités locales*, LGDJ, Collectivités locales, 1998, 131pp.

Rowe C. & Koetter F., Collage City, Infolio ed, Collection Archigraphy, Paris, 2002, 253pp.

Shahid Y. & Saich A., *China Urbanizes: Consequences, Strategies and Policies*, World Bank Publications, January 2008.

Smith N. & Willimans P., *Gentrification of the city*, Allen & Unwin ed, Boston, 1986.

Urry J., *Consuming places*, Routledge ed, London, 268pp.

Van Zeijl T., Koopmans B. & Doorn A., *De Gave Stad – Zoetermeer, 40 jaar lef en ambitie in architectuur*, Artned Publishing Support, Zoetermeer, 2006, 175pp.

Papers, Theses

Bécard L., "New Settlements: A New Approach to Solving Existing Housing and Urban Problems", in *The Expanding Metropolis: Coping with the Urban Growth of Cairo*, Cairo, 1985.

Burgel G., "La Ville Nouvelle comme politique urbaine: l'expérience des Villes Nouvelles de la région parisienne", in International Conference on New Towns Report, Ministry of Housing and Urban Development in Iran, Tehran, 2005.

Claval P., "De la logique des villes au ressort de l'urbanité", in Da Cunha A. & Mathey L., *La ville et l'urbain: des savoirs émergents*, Presses polytechniques et universitaires romandes, Months, 2007.

Gaborit P., "Development of New Towns and Peri-Urbanised Areas", in International Conference on New Towns Report, Ministry of Housing and Urban Development in Iran, Tehran, 2005, p. 275-290.

Helforoush R., "Fantasy city and identity in new towns", in *International Conference on New Towns Report, Ministry of Housing and Urban Development in Iran*, Tehran, 2005.

Hobson J., "New Towns, the modernist planning project and social justice: The case of Milton Keynes, UK, and 6th October, Egypt", University Coll, London, Development Planning Unit, 1999.

Hobson J., "New Towns, the modernist planning project and social justice" – Report from a conference on "Urban Development" organised in the town of 6th October, Cairo, Egypt, 1999.

Krier R., "Composition of urban spaces", in International Conference on New Towns Report, Ministry of Housing and Urban Development in Iran, Tehran, 2005.

Lapintie K., "New Towns, suburbs and urban sprawl – Struggling for integrity in the Helsinki metropolitan region", in International Conference on New Towns Report, Ministry of Housing and Urban Development in Iran, Tehran, 2005, p. 315-323.

Merlin P., "Les Villes Nouvelles à travers le monde – Essai de typologie", in International Conference on New Towns Report, Ministry of Housing and Urban Development in Iran, Tehran, 2005, p. 125-138.

Merlin P., "New Towns britanniques et Villes Nouvelles françaises: de la conception à l'achèvement", in International Conference on New Towns Report, Ministry of Housing and Urban Development in Iran, Tehran, 2005, p. 139-148.

Nguyen J., *The Challenge of Identity for European New Towns*, University Master Degree, University Lille I, 2009.

Oc T. & Tiesdell, "The death and life of city centres", in Oc T. & Tiesdell *Safer City Centres: Reviving the Public Realm*, Paul Chapman Publishing Ltd, London, 1997.

O'Connor J. & Wynne D., "From the margins to the centre: cultural production and consumption in the post-industrial city", in *Working Papers in Popular Cultural Studies*, No. 7, 1993.

Ruggieri D., *Constructing identity in master planned utopia: the new town of Irvine*, Ca, Conference communication paper, Alt Empordà, 2008.

Tayab Y., "Les rôles des différents acteurs dans l'aménagement et la gestion des villes nouvelles en France", in International Conference on New Towns Report, Ministry of Housing and Urban Development in Iran, Tehran, 2005.

Watson A. G., "New Towns in perspective in England", in *New Towns in Perspective*, INTA Press, London, 1991, p. 11.

Webster Douglas R. *et al.*, "Emerging Third Stage Peri-Urbanization: Functional Specialization in the Hangzhou Peri-Urban Region", The Urban Dynamics of East Asia Series, Asia/Pacific Research Center, Stanford: Stanford University, 2002.

Ziobrowski Z., *Problems of New Towns growth in Poland*, 2002, ENTP conference.

Studies, Reports

ACTED, Culture as a tool for development: challenges of analysis and action, Paris 2009.

American Chambre of Commerce in Egypt, Business Studies Series Report on New Communities, Cairo, 1995.

Bennett J., From New Towns to Growth Areas: learning from the past, Institute for Public Policy Research, London, 2005.

Boje Growth N., "Urban and regional identity challenged by globalisation", in European Cities in a Global Era, Danish Minister of the Environment, Copenhagen, 2002.

Communauté d'agglomération Cergy-Pontoise Énergies Ouest, Synthèse – Centre national de conservation, de restauration et de recherche du patrimoine, 2009.

Czischke D., Urban regeneration in Europe: the place of social housing in integrated urban policies, CECODHAS, Brussels, 2009.

Etablissement Public d'Aménagement de la Ville Nouvelle de Sénart, "Sénart poursuit sa tertiarisation", Press release, 2008.

European New Towns Platform, The top 8 specific challenges for social cohesion in New Towns, Brussels, 2005, 5pp.

European New Towns Platform, Contextual analysis of six New Towns, NEWTASC INTERREG IIIB programme, Brussels, 2006, 98pp.

European New Towns Platform, The top 8 specific challenges for urban regeneration in new Towns, Brussels, 2005, 11pp.

European Union, Territorial Agenda of the European Union – Towards a More Competitive and Sustainable Europe of Diverse Regions, Informal Minis–terial Meeting on urban Development and Territorial Cohesion, Leipzig, 2007.

Fouchier V., Les politiques des villes nouvelles (1965-2000), Programme interministériel d'histoire et d'évaluation des villes nouvelles françaises, Paris, 2000.

Goodbody Economic Consultants and UCD, "Development restraint and urban growth management", November 2000.

IAURIF – Institut d'Aménagement et d'Urbanisme d'Île de France, "Les temps de transport des actifs des Villes Nouvelles proches de ceux de la grande couronne", in Note rapide Population et modes de vie No. 398, Paris, 2005.

IAURIF – Institut d'Aménagement et d'Urbanisme d'Île de France, "Les villes nouvelles adoptées par leurs habitants", in Note rapide Population et modes de vie No. 394, Paris, 2005.

Imbert C., "L'ancrage residentiel et familial en ville nouvelle", in Note rapide Population et modes de vie No. 376/D, IAURIF, Paris, 2005.

International New Towns Institute, Research on New Towns – First International Seminar, D.H. Frieling Editor, 2006.

Interservice Group on Urban Development, Guide to the urban dimension in European Union policies, European Commission, Brussels, January 2010.

Partners in Regeneration, Involving Communities in Urban and Rural Regeneration – A guide for Practitioners, Department of the Environment, November 1995.

Social and Cultural Planning Office, New Towns Development, The Hague, April 2001.

Statistisch jaaroverzicht 2004, Gemeente Spijkenisse, the Netherlands.

Swedish Environmental Advisory Council, A strategy for energy-efficient buildings – Memorandum, Stockholm, 2004.

Town and Country Planning Association, The New Towns: their problems and future – Memorandum, 2002.

United Nations, Report of the World Commission on Environment and Development, 1987, General Assembly Resolution 42/187.

United Nations, "Millennium Ecosystem Assessment (MEA)", in Ecosystems and Human Well-Being: Synthesis, 2005, Island Press, Washington, 155pp.

United Nations Human Settlements Programme (UN-HABITAT), Promoting the Positive Rural-Urban Linkages Approach to Sustainable Development and Employment Creation: the Role of UN-HABITAT, Monitoring and Reasearch Division, 2003.

Articles

Antoni J. P., "Urban Sprawl as modelling: A methodological approach", in *Cybergeo*, No. 207, March 2002, p. 14.

ARTE Info, "Ordos, ville fantôme", Fried A. & Rambaud M., 5th of January 2010.

Baudelle G. & Ducom E., "Renouveler les Villes Nouvelles", in *Revue Urbanisme*, Paris, No. 362, 2008.

Bayou C and Le Bourhis C. "Les villes nouvelles de l'Est: quelles spécificités?", in *Regard sur l'Est, dossier sur les villes nouvelles*, November 2007, p. 1.

BBC News, "New Towns suffer 'major problems'", July 2002.

Beaufils S., Les temps de transport des actifs de villes nouvelles proches de ceux de grande couronne, IAURIF, 2005.

Berroir S. *et al.*, "La contribution des villes nouvelles au polycentrisme francilien: l'exemple de la polarisation liée à l'emploi", in *Espaces et sociétés* 1/2005 No. 119, p. 113-133.

Champenois M., "Les rides des New Towns", in *Le Monde*, Paris, 1976.

Chatterjeed Sudeshna "Naya Raipur, Chhattisgarh: A sustainable urban perspective", in *Sustainable solutions review*, p. 63-64.

Desné J., "Près de Shanghai, la ville nouvelle de Lingang attend ses résidents", *Agence France Presse* (AFP), 12th January 2009.

Donzelot F., "La ville à trois vitesses: relégation, péri-urbanisation, gentrification", in *Revue Esprit*, Paris, 2004.

Dupont V., "L'urbanisation en Inde: révision de quelques myths et vrais défis", in *Rayonnement du CNRS*, No. 47, Paris, mars 2008, 14pp.

Emelianoff C., "Les villes européennes face au développement durable: une floraison d'initiatives sur fond de désengagement politique", in *Cahiers du PROSES*, No. 8, Paris, Presses de Sciences-Po, 2004.

Gaborit P. & Van Leuwen L., "Les Villes Nouvelles en Europe: de la ville pilote à la ville durable", in *Pouvoirs locaux*, No. 60, Paris, 2004.

Gaborit P., "Reinventing the New Towns mindset", in *Nova Terra*, Delft, 2005.

Harlow Renaissance Ltd, Learning from Dutch New Towns and suburbs: report of the Harlow renaissance study tour, 2007.

Kroll L., "Quelle place pour la nature en ville?", in *La ville autrement*, *Alternatives Économiques*, Paris, 2009.

Molster A., "New Towns revisited", in 'SCAPE, 2006.

Nicolov Vlady, "Dimitrovgrad: quand l'identité cherche à s'émanciper du symbole", in *Regards sur l'Est, dossier sur les villes nouvelles*, November 2007.

Paquot T., "Éditorial – Villes indiennes", in *La Revue de l'Urbanisme* No. 355, July – August 2007.

Peiser R.B., Chang A.C. "Is it possible to build financially successful New Towns?", *in Urban Studies*, Vol. 36, No. 10, 1999, p. 1679-1703.

Ravenscroft N., "The Vitality and Viability of Town Centres", in No. 37, 2000.

Scheffler N., Kulikauskas P., Barreiro F., "Managing urban identities: aim or tool of urban regeneration?", in *URBACT Tribune*, 2009, p. 9.

Xu Shousong in Laowang Newspaper China, in *Courrier International*, France, July 2004.

Websites

New Towns

English Partnerships – The National Regeneration Agency.

http://www.englishpartnerships.co.uk/newtowns.htm.

The European New Towns Platform

www.newtowns.net.

European projects

CIUDAD – "From Pilot Towns to Sustainable Towns: Reinventing New Towns" http://www.ciudad-pilot.eu.

EGENIUS URBACT Project, Basline study, 2008.

http://urbact.eu/thematic-poles/sustainable-urban-development/thematic-networks/egenius/baseline.html.